T0181145

Ellipses Inscribed in, and Circumscribed about, Quadrilaterals

The main focus of this book is disseminating research results regarding the pencil of ellipses inscribing arbitrary convex quadrilaterals. In particular, the author proves that there is a unique ellipse of maximal area, EA, and a unique ellipse of minimal eccentricity, EI, inscribed in Q. Similar results are also proven for ellipses passing through the vertices of a convex quadrilateral along with some comparisons with inscribed ellipses. Special results are also given for parallelograms.

Researchers in geometry and applied mathematics will find this unique book of interest. Software developers, image processors along with geometers, mathematicians, and statisticians will be very interested in this treatment of the subject of inscribing and circumscribing ellipses with the comprehensive treatment here.

Most of the results in this book were proven by the author in several papers listed in the references at the end. This book gathers results in a unified treatment of the topics while also shortening and simplifying many of the proofs.

This book also contains a separate section on algorithms for finding ellipses of maximal area or of minimal eccentricity inscribed in, or circumscribed about, a given quadrilateral and for certain other topics treated in this book.

Anyone who has taken calculus and linear algebra and who has a basic understanding of ellipses will find it accessible.

Alan Horwitz holds a Ph.D. in Mathematics from Temple University in Philadelphia, PA, USA and is Professor Emeritus at Penn State University, Brandywine Campus where he served for 28 years. He has published 43 articles in refereed mathematics journals in various areas of mathematics. This is his first book.

Ellipses Inscribed in, and Circumscribed about, Quadrilaterals

Alan Horwitz

CRC Press
Taylor & Francis Group
Boca Raton London New York

CRC Press is an imprint of the
Taylor & Francis Group, an **informa** business

A CHAPMAN & HALL BOOK

First edition published 2024
by CRC Press
2385 Executive Center Drive, Suite 320, Boca Raton, FL 33431

and by CRC Press
4 Park Square, Milton Park, Abingdon, Oxon, OX14 4RN

CRC Press is an imprint of Taylor & Francis Group, LLC

© 2024 Alan Horwitz

Library of Congress Cataloging-in-Publication Data
Names: Horwitz, Alan, author.
Title: Ellipses inscribed in, and circumscribed about, quadrilaterals / authored by Alan Horwitz.
Description: First edition. | Boca Raton : CRC Press, 2024. |
Includes bibliographical references and index.
Identifiers: LCCN 2023054254 | ISBN 9781032622590 (hardback) |
ISBN 9781032756295 (paperback) | ISBN 9781003474890 (ebook)
Subjects: LCSH: Ellipse. | Quadrilaterals.
Classification: LCC QA485 .H77 2024 | DDC 516/.154–dc23/eng/20240316
LC record available at https://lccn.loc.gov/2023054254

ISBN: 9781032622590 (hbk)
ISBN: 9781032756295 (pbk)
ISBN: 9781003474890 (ebk)

DOI: 10.1201/9781003474890

Typeset in CMR10
by codeMantra

Contents

Preface

The main focus of this book is on ellipses inscribed in, or circumscribed about (passing through the vertices of), convex quadrilaterals, though more of the book focuses on inscribed ellipses. Most of the results in this book were proven by the author in several papers listed in the references at the end. The main motivation for writing this book was to give a unified presentation of the results, simplify or shorten certain proofs and/or provide new proofs, and compare inscribed and circumscribed ellipses. Some new results were also added, particularly for circumscribed ellipses.

My interest in ellipses inscribed in quadrilaterals started with completely determining the locus of centers of ellipses inscribed in Q. Of course, there is a well-known partial answer to this question – Newton's Theorem: If E is an ellipse inscribed in Q, then the center of E must lie on the open line segment, Z, connecting the midpoints of the diagonals of Q. However, what about the converse? That is, if $P \in Z$, must there be an ellipse with center P and inscribed in Q? We do answer that question in the book. While much is known about ellipses inscribed in triangles, somewhat less seems to be known for quadrilaterals. For example, it is known that there is a unique ellipse of maximal area and a unique ellipse of minimal eccentricity inscribed in any given triangle. The latter is trivial since every triangle contains a unique inscribed circle, which is then the ellipse of minimal eccentricity. Not every quadrilateral contains an inscribed circle, however, so the question about existence or uniqueness is not trivial in that case for minimal eccentricity. Of course, similar questions can be asked about circumscribed ellipses, with maximal area replaced by minimal area. We also answer those questions in this book, as well as provide some algorithms for finding such ellipses (that's a hint that such ellipses do exist and are unique, but that should not be surprising). One question we do not answer in this book is the question of maximizing or minimizing the arc length of ellipses inscribed in, or circumscribed about, convex quadrilaterals. The best we can do is for ellipses of minimal arc length inscribed in rectangles, a slight generalization of the known result for squares. As far as we know, this question has not been dealt with for triangles either.

Now the ellipse of minimal area inscribed in a triangle is the midpoint ellipse, which is tangent at the midpoint of all three sides. Is there an analog of the midpoint ellipse for convex quadrilaterals? It is well known that there is an ellipse inscribed in any given parallelogram which is tangent at the midpoint of all four sides. A recent paper proved that this cannot happen for quadrilaterals which are **not** parallelograms. So if Q is **not** a parallelogram, it is natural to ask – what is the **maximum** number of sides at which an ellipse inscribed in Q can be tangent to Q at the midpoint? Is there anything special we can say about a quadrilateral where that maximum occurs? The answer to these questions leads to a class of quadrilaterals, F, we call midpoint diagonal quadrilaterals, where the intersection point of the diagonals coincides with the midpoint of at least one of the diagonals. We recently discovered that these were also called bisect diagonal quadrilaterals. Of course, for a parallelogram, the intersection point of the diagonals lies on both diagonals, so in a certain sense the class F is a generalization of parallelograms. The midpoint diagonal quadrilaterals also have certain properties which parallelograms have with respect to tangency chords and conjugate diameters. More specifically, each ellipse inscribed in a parallelogram has a pair of parallel tangency chords and a pair of conjugate diameters which are parallel to the diagonals. We show that every midpoint diagonal quadrilateral also has these properties and that F is the largest class of quadrilaterals with these properties.

For circumscribed ellipses, we discuss Steiner's nice characterization of the ellipse of minimal eccentricity circumscribed about a convex quadrilateral, Q.

It is also interesting to compare inscribed and circumscribed ellipses for the same given quadrilateral, Q. For example, can an inscribed ellipse and a circumscribed ellipse have the same center? Is there a quadrilateral, Q, where the unique ellipses of minimal eccentricity inscribed in, or circumscribed about, Q have the **same** eccentricity? In addition to answering these questions, we also prove a relationship between the family of ellipses inscribed in a given parallelogram and the family of ellipses circumscribed about that same parallelogram.

One note about the methods used in the book. The approach we use is to derive equations for the family of ellipses inscribed in, or circumscribed about, a general convex quadrilateral, Q, which is not a parallelogram. More specific equations are given for parallelograms. Those equations are then used to provide formulas for the area or the eccentricity of inscribed or circumscribed ellipses. Thus we "get our hands dirty" with this approach. Other approaches, such as using projective geometry, are also very useful for proving certain results – particularly in relation to area. However, such an approach does not appear to work as well with proving results about eccentricity. Anyone with a basic understanding of ellipses and some knowledge of calculus should be able to understand and read most of this book.

Acknowledgement: I would like to thank CRC Press for giving me the opportunity to publish this book.

Part I

Ellipses Inscribed in Quadrilaterals

The main focus of Part 1 of this book is on ellipses inscribed in quadrilaterals, Q, in the xy plane. In Part 2 we discuss ellipses circumscribed about quadrilaterals and in Part 3 we give a comparison between the two. We assume throughout the book that all quadrilaterals are convex. By an *inscribed* ellipse, we mean an ellipse that lies inside Q and is tangent to each side of Q. We list here a series of questions one might ask about ellipses inscribed in quadrilaterals and then refer the reader to the section where those questions are answered.

1. Given a convex quadrilateral, Q, what is the locus of centers of ellipses inscribed in Q? A *partial* solution was given by Newton: If E is an ellipse inscribed in Q, then the center of E must lie on the open line segment, Z, connecting the midpoints of the diagonals of Q. However, this does not really give the precise locus of centers of ellipses inscribed in Q. One still must answer the following question: If $(x_0, y_0) \in Z$, must there be an ellipse inscribed in Q with center (x_0, y_0)? This question is answered fully in § 1.1.

2. Given a convex quadrilateral, Q, does there exist an ellipse of maximal area inscribed in Q and is such an ellipse unique? What about an ellipse of minimal eccentricity inscribed in Q? These questions are answered fully in § 1.2 and in § 1.3.

3. Not surprisingly, if one restricts the class of convex quadrilaterals, then it is possible to prove stronger results. In particular, for parallelograms, one can prove the following characterizations:

    (i) There is a unique ellipse, E_A, of maximal area inscribed in a parallelogram, b. In addition, E_A is tangent to b at the midpoints of the four sides of b (for more equivalencies, see Theorem 2.2).

    (ii) There is a unique ellipse, E_I, of minimal eccentricity inscribed in a parallelogram, b. In addition, the diagonals of b are equal conjugate diameters of E_I (see Theorem 2.3).

DOI: 10.1201/9781003474890-1

4. For a parallelogram, þ, it is known (or is not difficult to show) that the following two properties hold:

    (P1) Each ellipse inscribed in þ has tangency chords which are parallel to one of the diagonals of þ.

    (P2) Each ellipse inscribed in þ has a pair of conjugate diameters which are parallel to the diagonals of þ.

    Can one or both of these properties hold for quadrilaterals which are **not** parallelograms? If yes (that's a hint that there is such a quadrilateral), how does one characterize such quadrilaterals? These questions are answered fully in § 4, where we define and prove properties about a class of quadrilaterals we call midpoint diagonal quadrilaterals (mdq's). mdq's are, in a certain sense, a generalization of parallelograms.

5. As stated in #3 above, there is an ellipse inscribed in any given parallelogram, þ, which is tangent to þ at the midpoint of all four sides of þ. A recent paper (see [1]) proved that this cannot happen for quadrilaterals which are **not** parallelograms. So if Q is not a parallelogram, what is the **maximum** number of sides at which an ellipse inscribed in Q can be tangent to Q at the midpoint? Is there anything special we can say about a quadrilateral where that maximum occurs? These questions are answered fully in § 5.

6. Suppose that we are given a point, P, in the interior of a convex quadrilateral, Q, in the xy plane. Is there an ellipse inscribed in Q and which also passes through P? If yes, how many such ellipses? These questions are answered in § 6 (see, in particular, Theorem 6.1).

The methods used here rely directly on the general equation of an ellipse inscribed in a general convex quadrilateral. See Proposition 1.1(i) and Proposition 1.3(i). In addition, we only consider the Euclidean plane and not the projective plane. Other authors have used projective geometry to prove certain results about inscribed ellipses. In particular, see [2] and [3] for ellipses of maximal area inscribed in quadrilaterals. For ellipses of minimal eccentricity inscribed in a convex quadrilateral, projective geometric techniques are less useful because of the lack of affine invariance. In addition, the methods we use translate easily to algorithms for finding the ellipse of maximal area or of minimal eccentricity inscribed in a given quadrilateral. Finally, the methods we use might generalize to results about certain classes of simple closed curves (not just ellipses) inscribed in quadrilaterals.

There are numerous places outside of pure mathematics where ellipses inscribed in quadrilaterals are mentioned and/or used. The list of references at the end of this book is certainly not meant to be exhaustive. We do not go into any detail about these applications, but briefly mention them here. The following are the papers or dissertations that this author is aware of. The references involving some type of applications to areas outside of pure mathematics are [4], [5], [6], [7], [8], [9], and [10]. A related area is finding ellipses tangent to three given lines – some applications of this are given in [11] and [12]. In [4], the authors discuss solar coronal mass ejections (CMEs), which are significant drivers of adverse space weather. An algorithm from [13] is used to inscribe an ellipse in a quadrilateral, which provides a slice through the CME that matches the observations from 3D information. In [5], the authors discuss development of detection algorithms for airport security checks, which involve the identification of liquids contained in bottles in the hand luggage of flight passengers. They use Newton's Theorem and its converse mentioned in #1 above to help obtain the family of ellipses inscribed in the quadrilateral formed by the boundary rays from the two X-ray sources at which detection of the liquid starts and ends. In [6], the authors discuss a novel approach to representing a human hand based on a parallel binocular vision system. The algorithm they construct involves looking at the family of ellipses inscribed in a quadrilateral and then applying various constrains based on known physical properties. In [7], the authors discuss the applicability of the Diametric Faults system, which measures the diametric yarn unevenness and classifies the faults of yarn based on their geometric dimensions. Their purpose is to understand the influence of different yarn spinning systems, yarn fineness and opening and cleaning system on yarn diametric unevenness and yarn faults. They estimate the yarn cross-sectional eccentricity using results from [14]. In [8] and [9], the authors discuss packing of ellipses in a generic triangle or quadrilateral. They briefly refer to the papers [14] and [15]. In [10], the authors briefly refer to the paper [13].

Most of the results in this part were proven by the author in several papers listed in the references at the end – see [13]–[18], [15], and [19]. However, here the proofs provided of some of these results are different in certain ways – sometimes shortened and/or simplified, and in certain cases we use a completely different approach. In particular: The proof of existence/uniqueness for the ellipse of minimal eccentricity in [13] is simplified and a completely different proof of the area inequality in [17] is given here. In addition, § General Results on Ellipses in the Appendix provides easily accessible useful information for some of the proofs provided here and is also of some interest in its own right.

1

Locus of Centers, Maximal Area, and Minimal Eccentricity

1.1 Locus of Centers

We say that an ellipse, E_0, is inscribed in a convex quadrilateral, Q, if E_0 lies inside Q and is tangent to each side of Q. We assume in this chapter that Q is **not** a parallelogram. See § 2 for the case when Q is a parallelogram. We define an affine transformation, $T : R^2 \to R^2$ to be the map $T(\hat{x}) = A\hat{x} + \hat{b}$, where A is an invertible 2×2 matrix. Note that affine transformations map lines to lines, parallel lines to parallel lines, and preserve ratios of lengths along a given line. Also, the family of ellipses, tangent lines to ellipses, and four-sided convex polygons are preserved under affine transformations.

Definition 1 *For any given convex quadrilateral, Q, let M_1 and M_2 be the midpoints of the diagonals of Q. The Newton line segment, $\overline{M_1M_2}$, of Q is the open line segment connecting M_1 and M_2.*

This author's interest in ellipses inscribed in convex quadrilaterals, Q, began with the problem of determining the precise locus of centers of ellipses inscribed in Q, often referred to in the literature as Newton's problem. A *partial* solution was given by Newton as described in the following theorem.

Theorem 1.1 *(Newton) If E is an ellipse inscribed in Q, then the center of E must lie on the Newton line segment, $\overline{M_1M_2}$.*

Remark 1.1 *If Q is a parallelogram, then the diagonals of Q intersect at the midpoints of the diagonals of Q, and thus $\overline{M_1M_2}$ is really just one point.*

The proof of Theorem 1.1 involves first using an orthogonal projection to map E to a circle, C, and then proving Theorem 1.1 for C. Affine invariance then allows one to obtain Theorem 1.1 for ellipses in general. We shall not provide the details here and refer the reader to the very nice exposition given in [20]. See also [21]. However, Theorem 1.1 does not really give the precise locus of centers of ellipses, E, inscribed in a convex quadrilateral, Q. It only gives a necessary condition–the center of E must lie on the Newton line segment. But what about the converse? That is, is **every point** of the Newton line

DOI: 10.1201/9781003474890-2

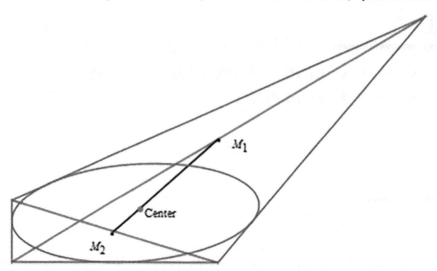

FIGURE 1.1
Illustration of Newton's Theorem

segment the center of some ellipse inscribed in Q? Newton's proof does not answer that question. Here is the converse of Newton's Theorem for convex quadrilaterals (along with a uniqueness result), which proves that the locus of centers of ellipses inscribed in Q is precisely the Newton line segment.

Theorem 1.2 *Let Q be a convex quadrilateral in the xy plane and let M_1 and M_2 be the midpoints of the diagonals of Q. If $(h_0, k_0) \in \overline{M_1 M_2}$, then there is a unique ellipse with center (h_0, k_0) inscribed in Q.*

Remark 1.2 *Again, as noted above for Theorem 1.1, if Q is a parallelogram, then the diagonals of Q intersect at the midpoints of the diagonals of Q, and thus $\overline{M_1 M_2}$ is really just one point.*

Before proving Theorem 1.2, we introduce some notation and also a set of quadrilaterals with a special set of vertices.

Notation 1 *We use the notation $Q(A_1, A_2, A_3, A_4)$ to denote the quadrilateral with vertices A_1, A_2, A_3, and A_4, starting with A_1 and going clockwise. Denote the sides of $Q(A_1, A_2, A_3, A_4)$ by S_1, S_2, S_3, and S_4, where $S_1 = \overline{A_1 A_2}, S_2 = \overline{A_2 A_3}, S_3 = \overline{A_3 A_4}$, and $S_4 = \overline{A_4 A_1}$. Denote the diagonals of $Q(A_1, A_2, A_3, A_4)$ by $D_1 = \overline{A_1 A_3}$ and $D_2 = \overline{A_2 A_4}$.*

Given a convex quadrilateral, $Q = Q(A_1, A_2, A_3, A_4)$, which is **not** a parallelogram, it will simplify our work below to consider quadrilaterals with a special set of vertices. Assume that $A_1 =$ lower left corner vertex. Then there is an affine transformation which sends A_1, A_2, and A_4 to the points $(0, 0), (0, 1)$,

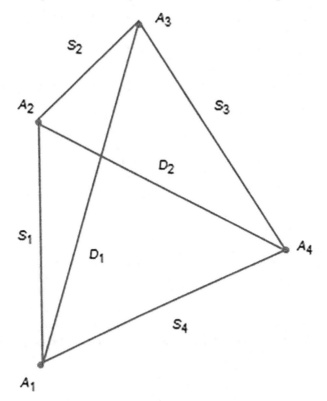

FIGURE 1.2
The quadrilateral $Q(A_1, A_2, A_3, A_4)$ with sides S_1, S_2, S_3, S_4 and diagonals D_1, D_2

and $(1, 0)$, respectively. It then follows that $A_3 = (s, t)$ for some $s, t > 0$. Thus it suffices to consider the quadrilateral, $Q_{s,t}$, with vertices $(0, 0), (0, 1), (s, t)$, and $(1, 0)$.

Also, if Q has a pair of parallel vertical sides, first rotate counterclockwise by $90°$, yielding a quadrilateral with parallel horizontal sides. Since we are assuming that Q is not a parallelogram, we may then also assume that $Q_{s,t}$ does not have parallel vertical sides and thus $s \neq 1$. Thus any trapezoid which is not a parallelogram may be mapped, by an affine transformation, to the quadrilateral $Q_{s,1}$. The sides of $Q_{s,t}$, going clockwise, are given by $S_1 = \overline{(0,0) \ (0,1)}, S_2 = \overline{(0,1) \ (s,t)}, S_3 = \overline{(s,t) \ (1,0)}$, and $S_4 = \overline{(0,0) \ (1,0)}$, and the corresponding lines which make up the boundary of $Q_{s,t}$ are given by

L_1: $x = 0$, L_2: $y = 1 + \dfrac{t-1}{s} x$, L_3: $y = \dfrac{t}{s-1}(x-1)$, and L_4: $y = 0$. Since $Q_{s,t}$ is convex and four-sided, (s, t) must lie above $\overline{(0,1) \ (1,0)}$, which implies that $s + t > 1$. Summarizing, we have

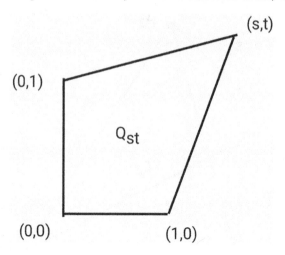

FIGURE 1.3
The quadrilateral $Q_{s,t}$

Lemma 1.1 *Suppose that Q is a convex quadrilateral which is **not** a parallelogram. Then there is an affine transformation which sends Q to the quadrilateral*

$$Q_{s,t} = Q(A_1, A_2, A_3, A_4),$$
$$A_1 = (0,0), A_2 = (0,1), A_3 = (s,t), A_4 = (1,0), \tag{1.1}$$
$$with \ (s,t) \in G, \ where$$

$$G = \{(s,t) : s, t > 0, s + t > 1, s \neq 1\}. \tag{1.2}$$

Remark 1.3 *All of the functions below, as well as the interval, I, depend on the parameters s and t. Except for the quadrilateral $Q_{s,t}$ itself, we suppress that dependence in our notation.*

The diagonals of $Q_{s,t}$ are $y = \dfrac{t}{s}x$ and $y = 1 - x$, and the midpoints of the diagonals of $Q_{s,t}$ are $M_1 = \left(\dfrac{1}{2}, \dfrac{1}{2}\right)$, $M_2 = \left(\dfrac{s}{2}, \dfrac{t}{2}\right)$. By Theorem 1.1, the x–coordinate of the center of any ellipse inscribed in $Q_{s,t}$ must lie in the open interval

$$I = \begin{cases} \left(\dfrac{s}{2}, \dfrac{1}{2}\right) & \text{if } s < 1 \\[2mm] \left(\dfrac{1}{2}, \dfrac{s}{2}\right) & \text{if } s > 1 \end{cases}. \tag{1.3}$$

The equation of the line thru M_1 and M_2 is

$$y = L(x) = \frac{1}{2}\frac{s - t + 2x(t-1)}{s-1}. \tag{1.4}$$

Throughout we let J denote the open interval $(0,1)$. Before proving Theorem 1.2, we need a lemma and and a couple of Propositions, which are useful in their own right.

Lemma 1.2 *If $(s,t) \in G$, then each of the following linear functions of q are positive on J. $y = (t-1)q+1, y = (t-1)q+s, y = (t-s)q+s, y = (t-1)(s+t)q+s, y = (s+t-2)q+1, y = qt, y = s(1-q), y = s^2(1-q), y = st(s+t-1)q$, and $y = 1-q$.*

Proof. It suffices to show that each of these linear functions of q are non–negative at the endpoints of J, which follows immediately from the fact that $q \in J$ and $(s,t) \in G$. ∎

Proposition 1.1 *Suppose that $(s,t) \in G$ and let J be the open line segment $(0,1)$.*

(i) E_0 is an ellipse inscribed in $Q_{s,t}$ if and only if the general equation of E_0 is given by

$$t^2 x^2 + (4q^2(t-1)t + 2qt(s-t+2) - 2st)xy +$$
$$((1-q)s + qt)^2 y^2 - 2qt^2 x - 2qt((1-q)s + qt)y + q^2 t^2 = 0 \tag{1.5}$$

for some $q \in J$.

(ii) If E_0 is an ellipse inscribed in $Q_{s,t}$ with equation given by (1.5) for some $q \in J$, then E_0 is tangent to the four sides of $Q_{s,t}$ at the points

$$\zeta_1 = \left(0, \frac{qt}{(t-s)q+s}\right) \in S_1,$$

$$\zeta_2 = \left(\frac{(1-q)s^2}{(t-1)(s+t)q+s}, \frac{t(s+q(t-1))}{(t-1)(s+t)q+s}\right) \in S_2,$$

$$\zeta_3 = \left(\frac{s+q(t-1)}{(s+t-2)q+1}, \frac{(1-q)t}{(s+t-2)q+1}\right), \zeta_4 = (q,0) \in S_4$$

The center of E_0 is given by

$$C_q = \left(\frac{1}{2}\frac{(t-s)q+s}{(t-1)q+1}, \frac{1}{2}\frac{t}{(t-1)q+1}\right).$$

The proof of Proposition 1.1 is given in the Appendix. One could prove Proposition 1.1 by showing that (1.5) yields an an ellipse inscribed in $Q_{s,t}$ and then by using uniqueness to show the converse. However, where does such an equation come from? That is answered in the proof given in the Appendix.

It is useful to note that by (1.5), the equation of E_0 can be written as $\Psi(x,y) = 0$, where $\Psi(x,y) = A(q)x^2 + B(q)xy + C(q)y^2 + D(q)x + E(q)y + F(q)$, and

$$
\begin{aligned}
A(q) &= t^2 \\
B(q) &= 4q^2(t-1)t + 2qt(s-t+2) - 2st \\
C(q) &= ((1-q)s + qt)^2 \\
D(q) &= -2qt^2 \\
E(q) &= -2qt((1-q)s + qt) \\
F(q) &= q^2 t^2.
\end{aligned}
\tag{1.6}
$$

Remark 1.4 *Note that $\overleftrightarrow{\zeta_1\zeta_3}$ and $\overleftrightarrow{\zeta_2\zeta_4}$ intersect at $\left(\dfrac{s}{s+t}, \dfrac{t}{s+t}\right)$, which is the intersection point of the diagonals of $Q_{s,t}$. This is not surprising since the same property holds for a circle inscribed in a tangential quadrilateral. We use this fact to simplify the proof of Proposition 1.1(ii) in the Appendix.*

Chakerian mentions the essence of Proposition 1.2 below in [20], but no proof is cited or given. We give a full proof in the Appendix.

Proposition 1.2 *Suppose that E_1 and E_2 are distinct ellipses with the same center and which are each inscribed in a convex quadrilateral, Q. Then Q must be a parallelogram.*

Propositions 1.1 and 1.2 imply the following:

Corollary 1 *(1.5) provides a one–to–one correspondence between ellipses inscribed in $Q_{s,t}$ and points $q \in J$.*

Proof. Suppose that E_1 and E_2 are ellipses inscribed in $Q_{s,t}$. Then the equations of E_1 and E_2 are given by (1.5) for some $q_1, q_2 \in J$ by Proposition 1.1. If $q_1 = q_2$, then $C_{q_1} = C_{q_2}$ and hence E_1 and E_2 have the same center. Thus $E_1 = E_2$ by Proposition 1.2. Conversely, if $E_1 = E_2$, then trivially E_1 and E_2 have the same center. As shown earlier in the proof of Proposition 1.2, the first component of C_q is a monotonic function of q (the second component is as well, but we don't need that here). That implies that $q_1 = q_2$. ∎

Theorem 1.2 was proven in [13] using a result of Marden relating the foci of an ellipse tangent to the lines thru the sides of a triangle and the zeros of a partial fraction expansion. There is a gap in [13] for the proof of uniqueness. We provide a more complete proof here.

Proof. Proof of Theorem 1.2. The case when Q is a parallelogram is easy to prove and we omit the details here. So assume that Q is **not** a parallelogram. Then by affine invariance, we may assume that $Q = Q_{s,t}$, the quadrilateral given in Lemma 1.1. Suppose now that $(h_0, k_0) \in \overline{M_1 M_2}$. Then $(h_0, k_0) = (h_0, L(h_0))$ for some $h_0 \in I$ (see (1.3)). If $f(q) = \dfrac{1}{2}\dfrac{(t-s)q + s}{(t-1)q + 1}$, then it follows easily that f is decreasing on J if $s > 1$ and increasing on J if $s < 1$. Hence f maps J onto I, which implies that $h_0 = \dfrac{1}{2}\dfrac{(t-s)q_0 + s}{(t-1)q_0 + 1}$ for some $q_0 \in J$. Let

E_0 be the conic with equation given by (1.5) with $q = q_0$. By Proposition 1.1, E_0 is an ellipse inscribed in $Q_{s,t}$, which proves existence. Uniqueness follows immediately from Corollary 1. ∎

1.2 Maximal Area

In this section we show that there is a unique ellipse of maximal area inscribed in any convex quadrilateral, Q, which is not a parallelogram. This result holds for parallelograms as well, but the proof is given later in § 2. As earlier, throughout we let J denote the open interval $(0, 1)$.

Theorem 1.3 *Let Q be a convex quadrilateral in the xy plane which is not a parallelogram. Then there is a unique ellipse of maximal area inscribed in Q.*

Proof. Let E_0 be an ellipse inscribed in Q. Then $(\text{area}(E_0))^2 = \pi^2 a^2 b^2$, where a and b are the lengths of the semi–major and semi–minor axes, respectively, of E_0. Maximizing the area amongst all ellipses inscribed in Q is equivalent to maximizing $a^2 b^2$ as a function of $q \in J$. By Lemma A.4, we may assume that $Q = Q_{s,t}$ for some $(s, t) \in G$ (see (1.2)). Using the equation of E_0 and the proof of Proposition 1.1 in the Appendix, $\Delta = 16t^2 (1 - q) q((t - 1)q + 1)((t - 1)q + s)$ and $\delta = 16t^4 q^2 (q - 1)^2 ((t - 1)q + s)^2$, which implies that $\dfrac{4\delta^2}{\Delta^3} = \dfrac{4(16t^4 q^2 (q - 1)^2 ((t - 1)q + s)^2)^2}{(16t^2 (1 - q) q((t - 1)q + 1)((t - 1)q + s))^3} = R(q)$, where, after simplifying, $R(q) = \dfrac{t^2 ((t - 1)q + s)(1 - q)q}{4((t - 1)q + 1)^3}$. By (A.2) of Lemma A.2,

$$a^2 b^2 = R(q), \tag{1.7}$$

so we want to maximize $R(q)$ as a function of $q \in J$. Now $R'(q) = -\dfrac{t^2 f(q)}{4((t - 1)q + 1)^4}$, where f is the polynomial

$$f(q) = (t - 1)(t - s + 2)q^2 + 2((s - 1)t + 1)q - s. \tag{1.8}$$

$f(0) = -s < 0$ and $f(1) = t(s + t - 1) > 0$, which implies that f has an odd number of roots in J. Since $\deg f \leq 2$, f must have precisely one root, q_0, in J, which implies that R has precisely one critical point, q_0, in J; By Lemma 1.2, $R(q) > 0$ on J. Since $R(0) = R(1) = 0$, $R(q_0)$ must be the unique global maximum of R on J. ∎

Some other authors have proven results about ellipses of maximal area inscribed in convex quadrilaterals. The following papers use, among other methods, projective geometry–see [22], [2], and [3]. In particular, an algorithm is given in [2] for finding the ellipse of maximal area inscribed in a general convex quadrilateral. In [22] the authors look at the special case of ellipses of maximal area inscribed in trapezoids and parallelograms (we consider the latter here in § 2.2). The methods (homogeneous coordinates in the projective plane) used in [2] and in [22] are somewhat different than those used above.

Remark 1.5 *For fixed s and t, if one lets $q \to 0^+$ or $q \to 1^-$, then the ellipses inscribed in $Q_{s,t}$ approach the line segments $ys - xt = 0$ or $x + y = 1$, respectively. Since a line segment has 0 area, that is why there is no ellipse of minimal area inscribed in $Q_{s,t}$.*

The following lemma will be useful in § 7 below in finding the unique ellipse of maximal area inscribed in a convex quadrilateral. We state it here since it directly relates to the proof of Theorem 1.3 above.

Lemma 1.3 *Let* $R(q) = \dfrac{t^2 \left((t-1)q + s\right)(1-q)q}{4\left((t-1)q + 1\right)^3}$ *and let* $h(s,t) = (st - t + 1)^2 + s(t-1)(t-s+2)$. *Suppose that $(s,t) \in G$ (see (1.2)).*
(i) If $t \neq 1$ and $t - s + 2 \neq 0$, then the unique critical point of R in J is given by $q = q_2 = \dfrac{-(st - t + 1) + \sqrt{h(s,t)}}{(t-1)(t-s+2)}.$
(ii) If $t - s + 2 = 0$, then the unique critical point of R in J is given by $q = \dfrac{1}{2}\dfrac{s}{(s-1)t + 1}.$ *This includes the case $t = 1$.*

Proof. By the proof of Theorem 1.3 above, the critical points of R are precisely the roots of f, where f is given by (1.8).
(i) If $t \neq 1$ and $t - s + 2 \neq 0$, then the roots of f are $q_1 = \dfrac{-(st - t + 1) - \sqrt{h(s,t)}}{(t-1)(t-s+2)}$ and q_2 above; Note that $q_2 - q_1 = \dfrac{2\sqrt{h(s,t)}}{(t-1)(t-s+2)}$; We consider the following cases.
Case 1: $(t-1)(t-s+2) > 0$; Since $\lim\limits_{q \to -\infty} f(q) = \infty$, f has a root in $(-\infty, 0)$. Since $q_2 > q_1$, q_2 must be the unique root of f in J.
Case 2: $(t-1)(t-s+2) < 0$; Since $\lim\limits_{q \to \infty} f(q) = -\infty$, f has a root in $(1, \infty)$. Since $q_2 < q_1$, again q_2 must be the unique root of f in J.
(ii) If $t - s + 2 = 0$, then it is easy to show that $(s-1)t + 1 \neq 0$ and thus the unique root of f in J is $q = \dfrac{1}{2}\dfrac{s}{(s-1)t + 1}.$ ∎

1.3   Minimal Eccentricity

In this section we show that there is a unique ellipse of minimal eccentricity inscribed in any convex quadrilateral, Q, when no two sides of Q are parallel

Locus of Centers, Maximal Area, and Minimal Eccentricity 13

(so that Q is **not** a trapezoid). Note that in the sections above, we used the weaker assumption that Q is **not** a parallelogram, but that will not suffice for this section. This was proven in [13], but the method here is somewhat different and a little shorter. Here we derive a formula for the eccentricity of an inscribed ellipse as a function of the coefficients of that ellipse's equation, which follows immediately from (A.5). Unfortunately, since the ratio of the eccentricity of two ellipses is **not** preserved in general under nonsingular affine transformations of the plane, we cannot assume, as in § 1.2, that $Q = Q_{s,t}$. However, by using an **isometry** of the plane, we can assume that Q has vertices $(0,0), (0,u), (s,t)$, and (v,w), where $s,v,u > 0$ and $t > w$. To obtain this isometry, first use a translation to map the lower left hand corner vertex of Q to $(0,0)$. Then a rotation about $(0,0)$, if necessary, yields a quadrilateral with vertices $(0,0), (0,u), (s,t)$, and (v,w) with $s,v > 0$. Finally, by using the map $T(x,y) = \left(\dfrac{1}{u}x, \dfrac{1}{u}y\right)$, we may assume that $u = 1$. By Lemmas A.5 and A.8, then, we may work with the quadrilateral $Q = Q_{s,t,v,w}$ for some s,t,v,w, where

$$
\begin{aligned}
Q_{s,t,v,w} &= Q(A_1, A_2, A_3, A_4), A_1 = (0,0), \\
A_2 &= (0,1), A_3 = (s,t), A_4 = (v,w),
\end{aligned}
\tag{1.9}
$$

and where

$$s, v > 0, t > w. \tag{1.10}$$

Letting $v = 1$ and $w = 0$ yields the quadrilateral $Q_{s,t}$–see (1.1). For example, let Q be the quadrilateral with vertices, going clockwise, $(-3,-1), (-1,4), (2,6)$, and $(4,1)$. Then the isometry $T(x,y) = \dfrac{1}{\sqrt{29}}(5x - 2y + 13, 2x + 5y + 11)$ maps Q to a quadrilateral with vertices $(0,0), (0,\sqrt{29}), \left(\dfrac{11}{\sqrt{29}}, \dfrac{45}{\sqrt{29}}\right)$, and $\left(\dfrac{31}{\sqrt{29}}, \dfrac{24}{\sqrt{29}}\right)$. Finally, the map $T(x,y) = \left(\dfrac{1}{\sqrt{29}}x, \dfrac{1}{\sqrt{29}}y\right)$ gives $Q_{s,t,v,w}$, where $s = \dfrac{11}{29}, t = \dfrac{45}{29}, v = \dfrac{31}{29}$, and $w = \dfrac{24}{29}$. See § 1.4 below for more details. Going clockwise, L_1: $x = 0, L_2$: $y = 1 + \dfrac{t-1}{s}x, L_3$: $y = w + \dfrac{t-w}{s-v}(x-v)$, and L_4: $y = \dfrac{w}{v}x$ denote the lines which make up the boundary of $Q_{s,t,v,w}$ and $S_1 = \overline{(0,0)\,(0,1)}, S_2 = \overline{(0,1)\,(s,t)}, S_3 = \overline{(s,t)\,(v,w)}$, and $S_4 = \overline{(0,0)\,(v,w)}$ denote the corresponding sides. The Newton Line has equation $L(x) = \dfrac{t}{2} + \dfrac{w+1-t}{v-s}\left(x - \dfrac{s}{2}\right)$. We now come up with some restrictions on s,t,v, and w arising from the fact that $Q_{s,t,v,w}$ is convex and that no two sides of $Q_{s,t,v,w}$ are parallel. We find it useful to define the following expressions, each of which depend on s,t,v, and w:

$$
\begin{aligned}
f_1 &= v(t-1) + (1-w)s, \tag{1.11} \\
f_2 &= vt - ws.
\end{aligned}
$$

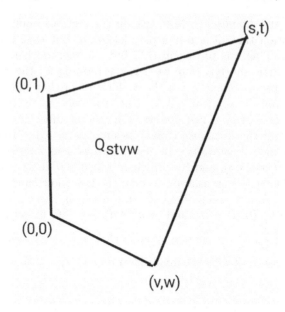

FIGURE 1.4
The quadrilateral $Q_{s,t,v,w}$

- Since $Q_{s,t,v,w}$ is convex and four-sided, (s,t) must lie above $\overleftrightarrow{(0,1)\,(v,w)}$ and (v,w) must lie below $\overleftrightarrow{(0,0)\,(s,t)}$, which implies that

$$f_1 > 0 \text{ and } f_2 > 0. \qquad (1.12)$$

- Since no two sides of Q are parallel, no two sides of $Q_{s,t,v,w}$ are parallel. In particular, $L_1 \not\parallel L_3$ and $L_2 \not\parallel L_4$, which easily implies that

$$f_1 \neq s \neq v. \qquad (1.13)$$

We now state a result somewhat similar to Proposition 1.1. Proposition 1.3(i) below gives necessary and sufficient conditions for the general equation of an ellipse inscribed in $Q_{s,t,v,w}$. It is useful for us to emphasize the dependence of the coefficients of the general equation on the parameter r in our notation.

Proposition 1.3 *Suppose that (1.10), (1.12), and (1.13) hold. Let J be the open line segment $(0,1)$.*

(i) E_0 *is an ellipse inscribed in* $Q_{s,t,v,w}$ *if and only if the general equation of* E_0 *is given by* $\psi(x,y) = 0$ *for some* $r \in J$, *where*

$$\psi(x,y) = A(r)x^2 + B(r)xy + C(r)y^2 + D(r)x + E(r)y + F(r), \qquad (1.14)$$

and

$$A(r) = (s^2 + v^2t^2 + w^2s^2 - 2tvs(w+1) + 2ws(2v-s))r^2$$
$$+ 2v\left(stw + st - 2ws - t^2v\right)r + t^2v^2,$$
$$B(r) = -2vs(2r^2(v-s) + rs(w+1) + v(t - rt - 2r)), \qquad (1.15)$$
$$C(r) = v^2s^2, D(r) = 2svr(-rs(w+1) + 2ws + tv(r-1)),$$
$$E(r) = -2v^2s^2r, F(r) = s^2v^2r^2.$$

(ii) If E_0 is an ellipse inscribed in $Q_{s,t,v,w}$ with equation given by $\psi(x,y) = 0$ for some $r \in J$, then E_0 is tangent to $Q_{s,t,v,w}$ at the points

$$\zeta_1 \;=\; (0,r), \zeta_2 = \frac{(sv(1-r), f_1r + vt(1-r))}{f_1r + v(1-r)},$$

$$\zeta_3 \;=\; \frac{((f_2 - (v-s)r)vs, wsf_1r + vtf_2(1-r))}{(s-v)(v+f_1)r + vf_2},$$

$$\zeta_4 \;=\; \frac{rs(v,w)}{rs + f_2(1-r)},$$

where f_1 and f_2 are given by (1.11). Also, the center of E_0 is

$$\frac{(sv, rs(1+w) - tv(r-1))}{2((s-v)r+v)}.$$

The proof of Proposition 1.3 is given in the Appendix. The proof of the theorem below requires looking at the following class of quadrilaterals.

Definition 2 *A convex quadrilateral, Q, is called tangential if there is a circle inscribed in Q.*

Now it is well known that Q is tangential if and only if $a+c = b+d$, where a and c, and b and d denote the lengths of opposite sides of Q. See, for example, [23]. For the quadrilateral $Q_{s,t,v,w}$, $a + c = \sqrt{v^2 + w^2} + \sqrt{s^2 + (t-1)^2}$ and $b+d = 1 + \sqrt{(s-v)^2 + (t-w)^2}$; For example, if $s = 2$, $t = 4$, $v = 1$, and $w = 2 - \frac{1}{6}\sqrt{66 + 6\sqrt{13}}$, then it can be shown that $\sqrt{v^2 + w^2} + \sqrt{s^2 + (t-1)^2} = 1 + \sqrt{(s-v)^2 + (t-w)^2}$ and thus $Q_{s,t,v,w}$ is tangential with these values of s, t, v, and w.

Theorem 1.4 *Let Q be a convex quadrilateral in the xy plane which is **not** a trapezoid. Then there is a unique ellipse of minimal eccentricity inscribed in Q.*

Remark 1.6 *We prove a similar result for trapezoids in § 1.5.*

Proof. Assume first that Q is tangential. Then by definition, there is a unique circle, Φ, inscribed in Q, and thus Φ is the unique ellipse of minimal eccentricity inscribed in Q since Φ has eccentricity 0. Now assume that Q is **not** tangential. This assumption will allow us to show that the function $G(r)$ below is differentiable on J. As discussed above, since Q is not a trapezoid, it suffices to assume that $Q = Q_{s,t,v,w}$ for some s, t, v, w, where (1.10), (1.12), and (1.13) hold. Let E_0 be an ellipse inscribed in $Q_{s,t,v,w}$ and let a and b denote the lengths of the semi–major and semi–minor axes, respectively, of E_0. Since the square of the eccentricity of E_0 equals $1 - \dfrac{b^2}{a^2}$, it suffices to maximize $\dfrac{b^2}{a^2}$, which is really a function of $r \in J$ since we allow E_0 to vary over all ellipses inscribed in $Q_{s,t,v,w}$. By Proposition 1.3, the general equation of E_0 is given by $\psi(x, y) = 0$, where ψ is given by (1.14) and (1.15). By (A.3) in Lemma A.2, $\dfrac{b^2}{a^2} = G(r)$, where $G(r) = \dfrac{O(r) - \sqrt{M(r)}}{O(r) + \sqrt{M(r)}}$ and

$$
\begin{aligned}
O(r) &= A(r) + C(r), &\qquad (1.16)\\
M(r) &= (A(r) - C(r))^2 + (B(r))^2.
\end{aligned}
$$

Also see [24]. Another useful formula is

$$
G(r) = \frac{N(r)}{(O(r) + \sqrt{M(r)})^2}, \qquad (1.17)
$$

where

$$
N(r) = O^2(r) - M(r). \qquad (1.18)
$$

After some simplification, N factors as

$$
N(r) = 16s^2 v^2 r\,(1 - r)\,((s - v)r + v)((s - v)r + f_2), \qquad (1.19)
$$

and N has roots

$$
r_1 = 0, r_2 = 1, r_3 = \frac{f_2}{v - s}, r_4 = \frac{v}{v - s}. \qquad (1.20)
$$

Note that $r_3 = r_4 \iff s - f_1 = 0$, which cannot hold by (1.13), and $r_3 \neq 0 \neq r_4$ since $v \neq 0 \neq f_2$ by (1.10) and (1.12). $r_3 = 1 \iff f_1 = 0$, which cannot hold by (1.12), and $r_4 = 1 \iff s = 0$, which cannot hold by (1.10). Thus all roots listed in (1.20) are *distinct*. Differentiating $G(r)$ using (1.17) and simplifying yields

$$
G'(r) = \frac{p(r)}{(O(r) + \sqrt{M(r)})^2 \sqrt{M(r)}}, \qquad (1.21)
$$

where p is the quartic polynomial

$$
p(r) = 2M(r)O'(r) - O(r)M'(r). \qquad (1.22)
$$

■

Before finishing the proof of Theorem 1.4, we state and prove the following lemmas. As usual, J denotes the open interval $(0, 1)$.

Lemma 1.4 *If* $s < v$, *then* $r_3 > 1$ *and* $r_4 > 1$. *If* $s > v$, *then* $r_3 < 0$ *and* $r_4 < 0$.

Proof. We give the details for r_3. If $s > v$, then $r_3 = \dfrac{f_2}{v - s} < 0$ by (1.12). If $s < v$, then $\dfrac{f_2}{v - s} - 1 = \dfrac{f_1}{v - s} > 0$ by (1.12), which implies that $r_3 > 1$. ∎

Lemma 1.5 $N(r) > 0$ *on* J.

Proof. First define the linear function of r, $F(r) = (s-v)r + f_2$. $F(0) = f_2 > 0$ and $F(1) = f_1 > 0$ by (1.12) and thus $F > 0$ on J. Similarly, $(s - v)r + v > 0$ on J. By (1.19), $N > 0$ on J since $r(1 - r) > 0$ on J. ∎

Lemma 1.6 *(i)* $O(r) > 0$ *on* J. *(ii)* $O(r_3) > 0$.

Proof. To prove (i), if $O(r_0) = 0$ for some $r_0 \in J$, then, by (1.18), $N(r_0) = -M(r_0)$. Since $M(r) \geq 0$ on J, $N(r_0)$ would not be positive, which contradicts Lemma 1.5. Hence $O(r)$ is nonzero on J. Since $O(0) = \left(s^2 + t^2\right) v^2 > 0$, that proves (i). To prove (ii), using computer algebra, one has $A(r_3) = \dfrac{(w(w - 1)s^2 + v(vt - 2ws)(t - 1))^2}{(s - v)^2} \geq 0$. Since $O(r_3) = A(r_3) + C(r_3)$ and $C(r) > 0$, that proves (ii). ∎

Lemma 1.7 *If* $Q_{s,t,v,w}$ *is not a tangential quadrilateral, then* $M(r) > 0$ *on* J *and thus* G *is differentiable on* J.

Proof. If $M(r_0) = 0$ for some $r_0 \in J$, then $A(r_0) - C(r_0) = 0$ and $B(r_0) = 0$, which implies that the ellipse inscribed in $Q_{s,t,v,w}$ corresponding to r_0 is a circle. But that contradicts the assumption that $Q_{s,t,v,w}$ is not a tangential quadrilateral. Thus $M(r) \neq 0$ for all $r \in J$, which implies that $M(r) > 0$ on J since M is non–negative. The differentiability of G on J then follows from Lemma 1.6 and (1.21). ∎

The following lemma finishes the proof of Theorem 1.4.

Lemma 1.8 p *has precisely one root in* J, *where* p *is defined by (1.22)*.

Proof. First, $p(0) = 16v^5 s^2 \left(s^2 + t^2\right) f_2 > 0$ and $p(1) = -16v^2 s^5 (v^2 + (w - 1)^2) f_1 < 0$, which implies that p has an odd number of roots in J. We shall find it useful to know the **signs** of $p(r_3)$ and of $p(r_4)$. First, $p(r_3) = 2M(r_3)O'(r_3) - O(r_3)M'(r_3)$ and, by (1.18), $M(r_3) = O^2(r_3)$ since $N(r_3) = 0$. Thus $p(r_3) = 2O^2(r_3)O'(r_3) - O(r_3)M'(r_3) = O(r_3)(2O(r_3)O'(r_3) - M'(r_3))$. Using computer algebra, one has $2O(r_3)O'(r_3) - M'(r_3) = -\dfrac{16v^2 s^2 f_1 f_2(s - f_1)}{s - v}$, which implies that

$$p(r_3) = -\frac{16v^2 s^2 f_1 f_2 (s - f_1)}{s - v} O(r_3). \tag{1.23}$$

Second, one can show that

$$p(r_4) = \frac{(16s^5v^5(s - f_1))((s - v)^2 + (t - w - 1)^2)}{(s - v)^3}. \qquad (1.24)$$

(1.23), and (1.24) then yield

$$(s - v)^4 p(r_3)p(r_4) = -16^2 v^7 s^7 f_1 f_2 (s - f_1)^2 O(r_3)\ x((s - v)^2$$
$$+ (t - w - 1)^2) < 0$$

by (1.12), (1.13), and Lemma 1.6(ii). Hence p has at least one root in (r_4, r_3) if $r_4 < r_3$ and in (r_3, r_4) if $r_3 < r_4$. Note that

$$r_3 - r_4 = \frac{s - f_1}{s - v}. \qquad (1.25)$$

We now consider the following cases. For each case, we prove that p has at least two roots outside J. Lemma 1.8 will then follow since $p(r_3)p(r_4) < 0$ implies that p has an *odd* number of roots *in* J and has no more than four roots total. Of course it then follows that p actually has exactly three roots outside J, though we don't need that fact here.

Case 1: $s > v$ and $s - f_1 > 0$. Then $r_4 < r_3 < 0$ by Lemma 1.4 and by (1.25), and $p(r_3) < 0$ by (1.23) and Lemma 1.6. Since $r_3 < 0, p(r_3) < 0$, and $p(0) > 0, p$ has at least one root in $(r_3, 0)$. The root in (r_4, r_3) yields at least two roots outside J.

Case 2: $s > v$ and $s - f_1 < 0$. Then $r_3 < r_4 < 0$ by Lemma 1.4 and by (1.25), and $p(r_4) < 0$ by (1.24). Since $r_4 < 0, p(r_4) < 0$, and $p(0) > 0, p$ has at least one root in $(r_4, 0)$. The root in (r_3, r_4) yields at least two roots outside J.

Case 3: $s < v$ and $s - f_1 > 0$. Then $1 < r_3 < r_4$ by Lemma 1.4 and by (1.25), and $p(r_3) > 0$ by (1.23) and Lemma 1.6. Since $r_3 > 1, p(r_3) > 0$, and $p(1) < 0, p$ has at least one root in $(1, r_3)$. The root in (r_3, r_4) yields at least two roots outside J.

Case 4: $s < v$ and $s - f_1 < 0$. Then $1 < r_4 < r_3$ by Lemma 1.4 and by (1.25), and $p(r_4) > 0$ by (1.24). Since $r_4 > 1, p(r_4) > 0$, and $p(1) < 0, p$ has at least one root in $(1, r_4)$. The root in (r_4, r_3) yields at least two roots outside J.

Note that G is differentiable on J by Lemma 1.7. Hence Lemma 1.8 and (1.21) imply that $G'(r)$ has precisely one root, $r_0 \in J$. Since $G(r) > 0$ on J, $G(0) = G(1) = 0$, and G is positive in the interior of I, $G(r_0)$ must yield the global maximum of G on J. That finishes the proof of Theorem 1.4. ∎

Remark 1.7 *For fixed s, t, v, and w, if one lets $r \to 0^+$ or $r \to 1^-$, then the ellipses inscribed in $Q_{s,t,v,w}$ approach the line segments $ys - xt = 0$ or $(1 - w)x + vy = v$, respectively. Since the eccentricity of those ellipses approaches 1, that is why there is no ellipse of maximal eccentricity inscribed in $Q_{s,t,v,w}$.*

1.4 Examples

Let Q be the quadrilateral with vertices $(-3,-1),(-1,4),(2,6)$, and $(4,1)$.

1. Find the general equation of an ellipse, E_Q, inscribed in Q.
   It is not hard to show that the affine transformation $\mathbb{F}_{s,t}(x,y) = \frac{1}{31}(5x - 2y + 13, -2x + 7y + 1)$ maps Q onto $Q_{s,t}$ (see 1.1), where $s = \frac{11}{31}$ and $t = \frac{39}{31}$. Let $E_{s,t} = \mathbb{F}_{s,t}(E_Q)$. By Proposition 1.1, after some simplifying, the equation of $E_{s,t}$ is $\Psi(x,y) = 0$, where $\Psi(x,y) = 1521x^2 + 78(16q^2 + 34q - 11)xy + (784q^2 + 616q + 121)y^2 - 3042qx - 78(28q^2 + 11q)y + 1521q^2, q \in J = (0,1)$. The equation

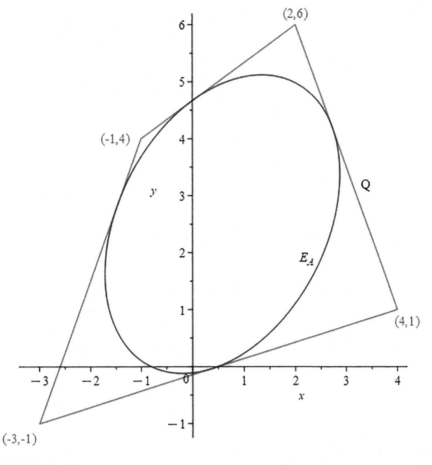

FIGURE 1.5
Ellipse of maximal area

of E_Q is then given by $\Psi(\mathbb{F}_{s,t}(x,y)) = 0$, which yields $1521(5x - 2y + 13)^2 + 78(16q^2 + 34q - 11)(5x - 2y + 13)(-2x + 7y + 1) + (784q^2 + 616q + 121)(-2x + 7y + 1)^2 - 94302q(5x - 2y + 13) - 2418(28q^2 + 11q)(-2x + 7y + 1) + 1461681q^2 = 0$.

2. Find the ellipse of maximal area, E_A, inscribed in Q. Since $t - s + 2 = \dfrac{90}{31} \neq 0$, by Lemma 1.3(i), E_A is obtained by letting $q = q_2 = \dfrac{1}{720}(-181 + \sqrt{m})$, where $m = 278\,281$; Substituting $q = q_2$ into the equation of E_Q given above yields the following approximate equation for E_A: $33347x^2 - 19604xy + 25534y^2 + 10517x - 116350y = 13085$.

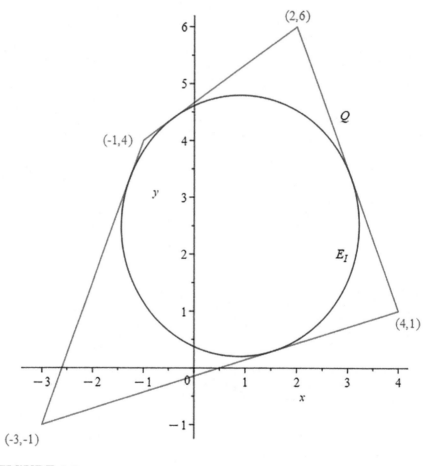

FIGURE 1.6
Ellipse of minimal eccentricity

3. Find the ellipse of minimal eccentricity, E_I, inscribed in Q.

Applying the affine transformation $\mathcal{F}_{s,t,v,w}(x,y) = \frac{1}{29}(5x - 2y + 13, 2x + 5y + 11)$ yields $\mathcal{F}_{s,t,v,w}(Q) = Q_{s,t,v,w}$ (see 1.9), where $s = \frac{11}{29}$, $t = \frac{45}{29}$, $v = \frac{31}{29}$, and $w = \frac{24}{29}$. Substituting those values for s, t, v, w into (1.15) and using (1.16), one has $O(r) = \frac{2}{24\,389}$ $(21\,928r^2 - 55\,428r + 35\,557)$ and $M(r) = \frac{4}{594\,823\,321}(666\,886\,784r^4 - 3268\,073\,568r^3 + 5845\,159\,492r^2 - 4504\,041\,708r + 1264\,300\,249)$. (1.22) then yields $\frac{14\,507\,145\,975\,869}{3720\,992}p(r) = 4872\,000r^4 - 42\,241\,296r^3 + 112\,474\,944r^2 - 118\,523\,974r + 42\,988\,413$. Since $s < v, s - f_1 < 0$, and $r_3 = \frac{39}{20} > r_4 = \frac{31}{20}$, we are in case 4 in the proof of Lemma 1.8 above. The unique root of p in J is $r_0 \approx 0.87$. Substituting $r = r_0$ along with the values above for s, t, v, w into (1.15) and using (1.14) yields $\psi(x,y) \approx 0.162\,80x^2 - 0.004145\,2xy + 0.164\,41y^2 - 0.136\,45 x - 0.284\,96y + 0.123\,48$. By Proposition 1.3, the ellipse, $E_{s,t,v,w}$ of minimal eccentricity inscribed in $Q_{s,t,v,w}$ has equation $\psi(x,y) = 0$. The equation of E_I is then given by $\psi(\mathcal{F}_{s,t,v,w}(x,y)) = 0$, which yields the approximate equation

$4.686\,2x^2 - 0.05484\,9xy + 4.\,802\,9y^2 - 8.\,250\,7x - 23.\,964y + 8.\,316\,8 = 0$. Using (A.3) of Lemma A.2, the minimal eccentricity for any ellipse inscribed in Q is ≈ 0.16375

1.5 Trapezoids

Throughout this section we discuss trapezoids which are **not** parallelograms. Theorems 1.2 and 1.3 from sections 1.1 and 1.2 each apply to such trapezoids, so there is no need to prove anything new here. However, in § 1.3 we assumed that the convex quadrilateral, Q, was not a trapezoid, mostly to make it easier to prove Theorem 1.4. Thus in this section we prove that there is a unique ellipse of minimal eccentricity inscribed in any trapezoid which is not a parallelogram (we handle parallelograms in the next section). To assume, as done earlier for non-trapezoids, that Q has a special form, use a rotation, if necessary, to map Q to a trapezoid with parallel vertical sides. Now use a translation to map the lower left hand corner vertex of Q to $(0,0)$. This isometry yields a quadrilateral with vertices $(0,0), (0, u), (s, t)$, and (s, w) with $s > 0$, which, by Lemma A.5, preserves the eccentricity of ellipses. In addition, by Lemma A.8, the map $T(x,y) = \left(\frac{1}{u}x, \frac{1}{u}y\right)$ also preserves the eccentricity of ellipses, and so we may assume that $u = 1$. That gives the quadrilateral $Q_{s,t,v,w}$ (see 1.9) with $v = s$, which we denote by $Q_{s,t,s,w}$. By Lemma A.5, this does not change the eccentricity of any inscribed ellipse. Proposition 1.4 below is then really just Proposition 1.3 with $v = s$.

Proposition 1.4 *Suppose that $s, t > 0$ and $t > w$.*

(i) *If E_0 is an ellipse inscribed in $Q_{s,t,s,w}$, then E_0 is tangent to $Q_{s,t,s,w}$ at the points $\zeta_1 = (0, r)$, $\zeta_2 = \dfrac{(s(1 - r), t - wr)}{(t - w - 1)r + 1}$,*

$\zeta_3 = (s, r(w - t) + t)$, *and* $\zeta_4 = \dfrac{r(s, w)}{r + (t - w)(1 - r)}$. *Also, the center of E_0 is given by $\dfrac{1}{2}(s, (1 + w - t)r + t)$.*

(ii) *E_0 is an ellipse inscribed in $Q_{s,t,s,w}$ if and only if the general equation of E_0 is given by $\psi(x, y) = 0$ for some $r \in J = (0, 1)$, where $\psi(x, y) = A(r)x^2 + B(r)xy + C(r)y^2 + D(r)x + E(r)y + F(r)$, and $A(r) = ((t - w - 1)^2 r^2 - 2(t^2 - t + 2w - wt)r + t^2)$, $B(r) = 2s(rt + r - rw - t)$, $C(r) = s^2$, $D(r) = 2rs(rt - rw - r + 2w - t)$, $E(r) = -2rs^2$, $F(r) = r^2 s^2$.*

Now we state and prove the analogy of Theorem 1.4 for trapezoids which are not parallelograms.

Theorem 1.5 *Let Q be a trapezoid in the xy plane which is not a parallelogram. Then there is a unique ellipse of minimal eccentricity inscribed in Q.*

Proof. Let E_0 be an ellipse inscribed in $Q_{s,t,s,w}$ and let a and b denote the lengths of the semi–major and semi–minor axes, respectively, of E_0. Since most of the details here are similar to the proof of Theorem 1.4, we shorten the proof. Using the notation from the proof of Theorem 1.4 for $G(r)$ and $p(r)$ (see (1.16), (1.17), and (1.22)) and the formulas for $A(r), B(r),$ and $C(r)$ given in Proposition 1.4, G is differentiable on J. The critical points of G are again the roots of $p(r) = 16s^2 (t - w) \alpha(r)$, where $\alpha(r) = 2(t - w - 1)^2 r^3 - 3(t - w - 1)^2 r^2 + 2(2w - t - wt - s^2)r + s^2 + t^2$. Now $\alpha(0) = s^2 + t^2 > 0$ and $\alpha(1) = -((w - 1)^2 + s^2) < 0$, which implies that α has an odd number of roots in $J = (0, 1)$. There are two cases to consider:

Case 1: $t - w - 1 \neq 0$. Then $\lim\limits_{r \to \infty} \alpha(r) = \infty \Rightarrow \alpha$ has at least one root in $(1, \infty)$. Thus α cannot have 3 roots in J, which implies that α has precisely one root, r_0, in J.

Case 2: $t - w - 1 = 0$. Then $\alpha(r) = -((t - 1)^2 + s^2 + 1)r + s^2 + t^2$. Since α is then a linear function, again it must have precisely one root, r_0, in J.

$G(r_0)$ must then yield the global maximum of G on J, which proves Theorem 1.5. ∎

Remark 1.8 *A different method of proof of Theorem 1.5 would involve considering a trapezoid with two parallel vertical sides as a limiting case of the quadrilaterals considered in § 1.3 and then using the results from 1.3 to show that there cannot be two ellipses of minimal eccentricity inscribed in Q. The method used here is perhaps longer, but has the advantage of providing direct formulas for the eccentricity of any ellipse inscribed in a trapezoid.*

2

Ellipses Inscribed in Parallelograms

In this chapter we prove numerous results about ellipses of maximal area (see Theorem 2.2) and of minimal eccentricity (see Theorem 2.3) inscribed in a parallelogram, the special case of ellipses inscribed in rectangles and a connection with maximal arc length (see Theorem 2.4), and so on. There are also some interesting connections with the foci of the ellipse of maximal area inscribed in a parallelogram, Q, the line of best fit for the vertices of Q (see Theorem 2.6), and tangency to Q at the midpoints of the four sides of Q (see Theorem 2.2 again).

2.1 Preliminary Results

We first state some preliminary results used to prove the main results in later subsections. The following lemma gives the equation and tangency points for an ellipse inscribed in a square.

Lemma 2.1 *Let S be the square with vertices $A_1 = (-1, -1), A_2 = (-1, 1)$, $A_3 = (1, 1)$, and $A_4 = (1, -1)$.*

  (i) *E_0 is an ellipse inscribed in S if and only if the general equation of E_0 is given by*

$$x^2 - 2uxy + y^2 + u^2 - 1 = 0, -1 < u < 1. \tag{2.1}$$

  (ii) *If the equation of E_0 is given by (2.1) for some $u \in (-1, 1)$, then E_0 is tangent to S at the points $\zeta_1 = (-1, -u) \in S_1, \zeta_2 = (u, 1) \in S_2, \zeta_3 = (1, u) \in S_3$, and $\zeta_4 = (-u, -1) \in S_4$.*

The proof of Lemma 2.1 is given in the Appendix. A similar version was proven in [14]. We simplified the proof a little bit here by choosing the vertices of Q so that the diagonals of Q intersect at the origin. Throughout we use the special parallelogram

$$Q_{d,k,l} = Q(A_1, A_2, A_3, A_4),$$
$$A_1 = (-l - d, -k), A_2 = (-l + d, k), A_3 = (l + d, k), \tag{2.2}$$
$$A_4 = (l - d, -k), l, k > 0; 0 \le d < l.$$

DOI: 10.1201/9781003474890-3

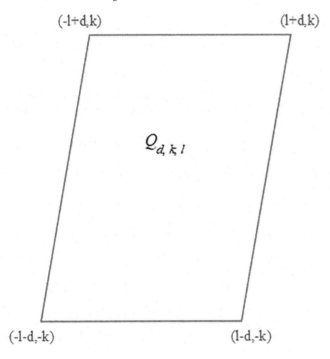

$(-1+d,k)$ $(1+d,k)$

$Q_{d,k,l}$

$(-1-d,-k)$ $(1-d,-k)$

FIGURE 2.1
The parallelogram $Q_{d,k,l}$

The diagonal lines of $Q_{d,k,l}$ are

$$\overleftrightarrow{D_1} : y = \frac{k}{d+l}x, \overleftrightarrow{D_2} : y = \frac{k}{d-l}x. \qquad (2.3)$$

One can easily prove the following fact, which we use later:

Lemma 2.2 $Q_{d,k,l}$ *is a rectangle if and only if* $d = 0$ *and* $Q_{d,k,l}$ *is a rhombus if and only if* $d^2 + k^2 = l^2$.

The following proposition gives the equation and tangency points for an ellipse inscribed in $Q_{d,k,l}$.

Proposition 2.1 *(i)* E_0 *is an ellipse inscribed in* $Q_{d,k,l}$ *if and only if the general equation of* E_0 *is given by*

$$k^2 x^2 - 2k(d + lu)xy + \qquad (2.4)$$
$$(d^2 + l^2 + 2dlu)y^2 + k^2 l^2 (u^2 - 1) = 0, -1 < u < 1.$$

(ii) If E_0 is the ellipse given by (2.4) for some u, $-1 < u < 1$, then E_0 is tangent to the four sides of $Q_{d,k,l}$ at the points $\zeta_1 = (-l - du, -ku), \zeta_2 = (lu + d, k), \zeta_3 = (l + du, ku)$, and $\zeta_4 = (-lu - d, -k)$.

Proof. Let $\mathbb{F}$ be the non–singular linear transformation given by $\mathbb{F}(x,y) = (lx+dy, ky)$, which maps the square, S, onto $Q_{d,k,l}$. Then $\mathbb{F}^{-1}(x,y) = \frac{1}{kl}(kx - dy, ly)$; By Lemma 2.1, replacing x by $\frac{kx - dy}{kl}$ and replacing y by $\frac{1}{kl}(ky) = \frac{y}{l}$ in (2.1) yields (2.4). Applying $\mathbb{F}$ to the points of tangency in Lemma 2.1 gives the points of tangency in Proposition 2.1(ii). The rest of Proposition 2.1 follows from Lemma 12.3 ∎

The following result is probably known–we prove it using Proposition 2.1.

Corollary 2 *A parallelogram is tangential if and only if it is a rhombus.*

Proof. As usual, we may work with the parallelogram $Q_{d,k,l}$ for some $l, k > 0$; $0 \le d < l$. If $Q_{d,k,l}$ is tangential, then (2.4) defines the equation of a circle only if the coefficient of xy equals 0 and the coefficient of x^2 equals the coefficient of y^2. That yields $d + lu = 0 \Rightarrow u = -\frac{d}{l}$, which in turn implies that $d^2 + l^2 + 2dlu = d^2 + l^2 + 2dl\left(-\frac{d}{l}\right) = l^2 - d^2$. One also requires $k^2 = l^2 - d^2$, and thus $Q_{d,k,l}$ is a rhombus. Conversely, if $Q_{d,k,l}$ is a rhombus, then $d^2 + k^2 = l^2$. Letting $u = -\frac{d}{l} \in (-1,1)$ in (2.4) yields the equation of the circle $k^2x^2 + k^2y^2 = k^2l^2\frac{l^2 - d^2}{l^2}$ or $x^2 + y^2 = k^2$ inscribed in $Q_{d,k,l}$. ∎

Letting $d = 0$ in Proposition 2.1 yields:

Corollary 3 *Let Z be the rectangle with vertices $(-l, -k), (-l, k), (l, k)$, and $(l, -k)$, where $l, k > 0$.*

(i) *E_0 is an ellipse inscribed in Z if and only if the general equation of E_0 is given by*

$$k^2x^2 - 2kluxy + l^2y^2 + k^2l^2(u^2 - 1) = 0, -1 < u < 1. \qquad (2.5)$$

(ii) *If E_0 is the ellipse given by (2.5) for some $u, -1 < u < 1$, then E_0 is tangent to the four sides of Z at the points $\zeta_1 = (-l, -ku), \zeta_2 = (lu, k), \zeta_3 = (l, ku)$, and $\zeta_4 = (-lu, -k)$.*

We find it useful to introduce the following notation:

$$\begin{aligned} J &= d^2 + k^2 + l^2 \\ I &= d^2 - k^2 + l^2 \\ N &= d^2 + k^2 - l^2. \end{aligned} \qquad (2.6)$$

The following lemma expresses the lengths of the axes of an ellipse inscribed in $Q_{d,k,l}$ using the equation of such an ellipse.

Lemma 2.3 *Let E_0 denote any ellipse inscribed in $Q_{d,k,l}$ and let a and b denote the lengths of the semi–major and semi–minor axes, respectively, of E_0. Suppose that the equation of E_0 is given by (2.4). Then $a^2 = \dfrac{p(u) + \sqrt{q(u)}}{2}$ and $b^2 = \dfrac{p(u) - \sqrt{q(u)}}{2}$, where*

$$p(u) = J + 2dlu, \tag{2.7}$$
$$q(u) = (2uld + I)^2 + 4k^2 (ul + d)^2.$$

Proof. By (2.4), the equation of E_0 is given by $Ax^2 + Bxy + Cy^2 + Dx + Ey + F = 0$, where $A = k^2$, $B = -2k(d + lu)$, $C = d^2 + l^2 + 2dlu$, $D = E = 0$, and $F = k^2 l^2 (u^2 - 1)$, $-1 < u < 1$. Some simplification yields $\Delta = 4AC - B^2 = 4k^2 l^2 (1 - u^2)$ and $\delta = CD^2 + AE^2 - BDE - F\Delta = 4k^4 l^4 (1 - u^2)^2$. Then $\mu = \dfrac{4\delta}{\Delta^2} = 1$, which implies that $a^2 = \dfrac{A + C + \sqrt{(A - C)^2 + B^2}}{2}$ and $b^2 = \dfrac{A + C - \sqrt{(A - C)^2 + B^2}}{2}$ by (12.5) in the proof of Lemma 12.2 in the Appendix. Some more simplification yields (2.7). ■

2.2 Maximal Area

Many of the results in this section were proven in [17] and were also known prior to the results in [17], though most of the proofs given here are somewhat different. Also, some of the results here hold for a more general class of quadrilaterals called the midpoint diagonal quadrilaterals (mdq's), which are discussed in § 4. Some of the results are known (see, for example, [1]), though again the proofs are different and several results are new. Our first main result is Theorem 2.2, which establishes three properties of the ellipse of maximal area, E_A, inscribed in Q related to tangency at the midpoints of the sides of Q, an area inequality involving E_A, and a connection between the critical points of the polynomial with roots at the vertices of Q and the foci and center of E_A. Before stating and proving Theorem 2.2, we now state the well-known Bocher–Grace theorem (see [25]).

Theorem 2.1 *The critical points of a cubic , p, are the foci of an ellipse, E_0, which is tangent to the midpoints of the sides of the triangle formed by joining the roots of p.*

Theorem 2.2 *Let Q be a parallelogram in the xy plane. Then there is a unique ellipse, E_A, of maximal area inscribed in Q. Furthermore*

(i) E_A is tangent to Q at the midpoints of the four sides of Q.

(ii) $\dfrac{Area\,(E_A)}{Area(Q)} = \dfrac{\pi}{4}.$

(iii) If P is a polynomial with roots at the vertices of Q, then the three roots of $P'(z)$ are the foci and center of E_A.

Remark 2.1 *(i) is the well-known result that E_A is the midpoint ellipse.*

Proof. Note that ratios of areas of ellipses and four–sided convex polygons are preserved under one–one affine transformations. Since every parallelogram is affinely equivalent to the square, S, with vertices $A_1 = (-1,-1), A_2 = (-1,1), A_3 = (1,1)$, and $A_4 = (1,-1)$, for the proof of existence/uniqueness and of (i) and (ii), we may assume that $Q = S$. So let E_0 denote an ellipse inscribed in S and let a and b denote the lengths of the semi–major and semi–minor axes, respectively, of E_0. Using Lemma 2.1(i) and letting $A = 1$, $B = -2u$, $C = 1$, $D = 0$, $E = 0$, $F = u^2 - 1$, we have $\Delta = 4AC - B^2 = 4 - 4u^2$ and $\delta = CD^2 + AE^2 - BDE - F\Delta = 4\left(u^2 - 1\right)^2$. By Lemma 12.2, $a^2 b^2 = \dfrac{4\delta^2}{\Delta^3} = \dfrac{64\left(u^2 - 1\right)^4}{64(1 - u^2)^3} = 1 - u^2$, which clearly attains its unique global maximum, for $-1 < u < 1$, if and only if $u = 0$. Letting $u = 0$ in Lemma 2.1(ii), E_A is tangent to the four sides of S at the points $(-1,0)\,, (0,1)\,, (1,0)$, and $(0,-1)$, which are the midpoints of the sides of S. That proves existence/uniqueness and (i). Again, with $u = 0$, we have $a^2 b^2 = 1$, which implies that $\dfrac{Area\,(E_A)}{Area(S)} = \dfrac{\pi}{4}$ and proves (ii). (iii) follows immediately from Theorem 2.1. ∎

Remark 2.2 *Note that the ellipse of maximal area inscribed in the square S is the unit circle $x^2 + y^2 = 1$.*

2.3   Minimal Eccentricity

The following lemma allows us to simplify the proof of the main result in this section, Theorem 2.3 below.

Lemma 2.4 *: Any parallelogram, Q, is isometric to the parallelogram, $Q_{d,k,l}$ for some $l, k > 0;\ 0 \leq d < l$.*

The proof of Lemma 2.4 is given in the Appendix.

The following theorem establishes an interesting connection between equal conjugate diameters of the ellipse, E_I, of minimal eccentricity inscribed in any parallelogram, Q, and the diagonals of Q. It also answers the question of when, if ever, $E_A = E_I$ among the class of parallelograms.

Theorem 2.3 *Let Q be a parallelogram in the xy plane. Then*

(i) There is a unique ellipse, E_I, of minimal eccentricity inscribed in Q.

(ii) The diagonals of Q are equal conjugate diameters of E_I.

(iii) If the unique ellipse of maximal area inscribed in Q equals the unique ellipse of minimal eccentricity inscribed in Q, then Q must be a rectangle.

Remark 2.3 *In [14], we proved that the smallest nonnegative angle between equal conjugate diameters of E_I equals the smallest nonnegative angle between the diagonals of Q. (ii) above is obviously a stronger result and follows immediately from the following fact: Let L and L' denote a pair of conjugate diameters of an ellipse inscribed in a parallelogram, Q. Then if L and L' are parallel to the diagonals of Q, they must actually **equal** those diagonals since the diameters and the diagonals both pass thru the center of Q.*

Remark 2.4 *In § 4 below, we define a new class of quadrilaterals we call midpoint diagonal quadrilaterals (mdq's). A parallelogram is a special case of an mdq and in a certain sense the mdq's generalize parallelograms. In § 4 we show that if Q is an mdq and E_I is the unique ellipse of minimal eccentricity inscribed in Q, then the equal conjugate diameters of E_I are parallel to the diagonals of Q. (ii) above would then follow from the remark above and the fact that a parallelogram is a special case of an mdq. However, we provide the details here, partly because (iii) would not follow from any result for mdq's in general.*

Remark 2.5 *The converse of (iii) is proven below in Theorem 2.4 in § 2.4.*

Proof. As in § 1.3 for convex quadrilaterals in general, by using an **isometry** of the plane, by Lemma 2.4 we may assume that $Q = Q_{d,k,l}$ for some $l, k > 0$; $0 \leq d < l$. Let E_0 denote an ellipse inscribed in $Q_{d,k,l}$ and let a and b denote the lengths of the semi–major and semi–minor axes, respectively, of E_0. By Lemma 2.3,

$$\frac{b^2}{a^2} = f(u) = \frac{p(u) - \sqrt{q(u)}}{p(u) + \sqrt{q(u)}}, -1 < u < 1, \tag{2.8}$$

where p and q are given by (2.7). Some simplification yields

$$f(u) = \frac{4k^2l^2\left(1 - u^2\right)}{\left(p(u) + \sqrt{q(u)}\right)^2}. \tag{2.9}$$

Now $p(u) + \sqrt{q(u)} = 0 \Rightarrow p^2(u) - q(u) = 0 \Rightarrow 4k^2l^2\left(1 - u^2\right) = 0$, which cannot occur, and thus f is a differentiable function of u on $(-1, 1)$ since p and q are differentiable functions. Using (2.8), $f' = 0$ implies that $(p + \sqrt{q})\left(p' - \frac{q'}{2\sqrt{q}}\right) - (p - \sqrt{q})\left(p' + \frac{q'}{2\sqrt{q}}\right) = 0$, where derivatives are taken with respect to u. Simplifying yields $pp' - \dfrac{pq'}{2\sqrt{q}} + \sqrt{q}p' - \dfrac{1}{2}q' - pp' - \dfrac{pq'}{2\sqrt{q}} + \sqrt{q}p' +$

$\dfrac{1}{2}q' = -\dfrac{pq'}{\sqrt{q}} + 2\sqrt{q}p' = 0$, which implies that $-pq' + 2qp' = 0$, and thus $-(J +$

$2dlu)(2\,(2uld + I)\,2dl + 8k^2 l\,(ul + d)\,) + 2(\,(2uld + I)^2 + 4k^2\,(ul + d)^2\,)(2dl) =$

$0 \Rightarrow -2l^2 k^2 (Ju + 2ld) = 0$. Hence the only critical point of f is

$$u = u_\epsilon = -\frac{2dl}{J}. \tag{2.10}$$

Note that $J - 2dl = k^2 + (d - l)^2 > 0$. Now $u_\epsilon < 0$, and $1 + u_\epsilon = 1 - \dfrac{2dl}{J} =$

$\dfrac{J - 2dl}{J} = \dfrac{k^2 + (d - l)^2}{J} > 0$, which implies that $-1 < u_\epsilon < 1$. Also, $p(-1) =$

$J - 2ld$ and $q(-1) = (J - 2ld)^2$, which implies that $p(-1) + \sqrt{q(-1)} = J -$
$2ld + |J - 2ld| = 2\,(J - 2ld) \neq 0$. By (2.9) we have $f(-1) = f(1) = 0$ and
$f(u) > 0$ for $-1 < u < 1$, which implies that f attains its global maximum at
u_ϵ and the eccentricity is minimized when $u = u_\epsilon$. That proves (i). To prove
(ii): By Proposition 2.1, the equation of E_I is given by (2.4) with $u = u_\epsilon$,
which yields $k^2 x^2 - 2k(d + lu_\epsilon)xy + (d^2 + l^2 + 2dlu_\epsilon)y^2 + k^2 l^2 (u_\epsilon^2 - 1) = 0$.
Substituting for u_ϵ using (2.10) and simplifying yields

$$J^2 k^2 x^2 - 2kdJ(J - 2l^2)xy$$
$$+ J((d^2 + l^2)J - 4d^2 l^2)y^2 + k^2 l^2 (4d^2 l^2 - J^2) = 0. \tag{2.11}$$

First we prove that the diagonal line segments of $Q_{d,k,l}, D_1$ and D_2, are conjugate diameters of E_I (see (2.3)). Differentiating (2.11) with respect to x and
simplifying yields

$$\frac{dy}{dx} = \frac{-k^2 Jx + kd(J - 2l^2)y}{-kd(J - 2l^2)x + ((d^2 + l^2)J - 4d^2 l^2)y}. \tag{2.12}$$

Now suppose that $\overleftrightarrow{D_1}$ intersects E_I at the two distinct points $P_j = (x_j, y_j) =$
$\left(x_j, \dfrac{k}{d+l}x_j\right), j = 1, 2$. By (2.11),

$$J^2 k^2 x_j^2 - 2kdJ(J - 2l^2)\left(\frac{k}{d+l}\right)x_j^2 + J((d^2 + l^2)J - 4d^2 l^2)\left(\frac{k}{d+l}\right)^2 x_j^2 +$$

$k^2 l^2 (4d^2 l^2 - J^2) = 0$. After simplifying we have $\dfrac{2Jk^2 l^2 (J + 2dl)}{(d+l)^2} x_j^2 = k^2 l^2 (J^2 -$

$4d^2 l^2) \Rightarrow x_j^2 = (d+l)^2\,\dfrac{(J^2 - 4d^2 l^2)}{2J(J + 2dl)} = \dfrac{(J - 2dl)\,(d+l)^2}{2J} \Rightarrow$

$$x_j = \pm(d+l)\sqrt{\frac{J-2dl}{2J}},$$

$$y_j = \pm k\sqrt{\frac{J-2dl}{2J}}, j = 1, 2. \tag{2.13}$$

Plugging (x_j, y_j) into (2.12) and simplifying gives $\dfrac{dy}{dx} = \dfrac{k}{d-l} = $ slope of $\overleftrightarrow{D_2}$, which proves that the tangent line at P_1 or at P_2 is parallel to $\overleftrightarrow{D_2}$. Hence D_1 and D_2 are conjugate diameters of E_I. Similarly, suppose that D_2 intersects E_I at the two distinct points $P_j = (x_j, y_j) = \left(x_j, \dfrac{k}{d-l}x_j\right), j = 3, 4$. As done above, one can show that

$$x_j = \pm(d-l)\sqrt{\frac{J+2dl}{2J}},$$

$$y_j = \pm k\sqrt{\frac{J+2dl}{2J}}, j = 3, 4. \tag{2.14}$$

To prove that D_1 and D_2 are **equal** conjugate diameters of E_I, one must show that $|P_1P_2|^2 = |P_3P_4|^2$ or that $(x_2-x_1)^2 + (y_2-y_1)^2 = (x_4-x_3)^2 + (y_4-y_3)^2$. By (2.13) and (2.14) this is equivalent to $((d+l)^2 + k^2)\dfrac{J-2dl}{J} = ((d-l)^2 + k^2)\dfrac{J+2dl}{J} \iff ((d+l)^2 + k^2)(J - 2dl) = ((d-l)^2 + k^2)(J + 2dl)$, which follows easily. That proves (ii). To prove (iii): In the proof of Theorem 2.2 we showed that $u = 0$ corresponds to the unique ellipse of maximal area inscribed in $Q_{d,k,l}$, while in the proof of (i) above, we showed that $u = u_\epsilon = -\dfrac{2dl}{J}$ corresponds to the unique ellipse of minimal eccentricity inscribed in Q. Now $-\dfrac{2dl}{J} = 0 \iff d = 0$, which proves (iii). ∎

2.4 Special Result for Rectangles

Definition 3 *A rectangle is a parallelogram all of whose angles are right angles.*

Theorem 2.4 *Let Z be a rectangle in the xy plane, let E_A be the unique ellipse of maximal area inscribed in Z, and let E_I be the unique ellipse of minimal eccentricity inscribed in Z. Then*

(A) $E_A = E_I$.

(B) E_A is the unique ellipse of maximal arc length inscribed in Z.

Proof. Let E_0 denote an ellipse inscribed in Z and let a and b denote the lengths of the semi–major and semi–minor axes, respectively, of E_0. By Lemma 12.3 , using a translation and/or a rotation, we may assume that the vertices of Z are $(-l, -k), (-l, k), (l, k)$, and $(l, -k)$, where $l, k > 0$. By Lemma 2.3 with $d = 0$, $\dfrac{b^2}{a^2} = f(u)$, where $f(u) = \dfrac{k^2 + l^2 - \sqrt{g(u)}}{k^2 + l^2 + \sqrt{g(u)}}$ and

$$g(u) = (l^2 - k^2)^2 + 4k^2 l^2 u^2.$$

To prove (A), we minimize the eccentricity by maximizing $f(u)$ as a function of u. Some simplification yields

$$f(u) = \frac{4k^2 l^2 \left(1 - u^2\right)}{(k^2 + l^2 + \sqrt{g(u)})^2}.$$

Differentiating with respect to u and simplifying yields

$$f'(u) =$$
$$\frac{-8k^2 l^2 u((k^2 + l^2 + \sqrt{g(u)})^{3/2} + 4k^2 l^2 \left(1 - u^2\right))}{(k^2 + l^2 + \sqrt{g(u)})^{7/2}}.$$

Note that $k^2 + l^2 + \sqrt{g(0)} = l^2 + k^2 + \left|l^2 - k^2\right| > 0$, which implies that $f'(0)$ exists. Also, since $(k^2 + l^2 + \sqrt{g(u)})^{3/2} + 4k^2 l^2 \left(1 - u^2\right) > 0$ for $-1 < u < 1$, $f'(u) = 0 \iff u = 0$. Since $f(-1) = f(1) = 0$ and $f(u) > 0$ for $-1 < u < 1$, f attains its global maximum when $u = 0$ and thus E_I occurs $\iff u = 0$. As done in the proof of Theorem 2.2 where we used a square, it is not hard to show that $u = 0$ corresponds to the unique ellipse of maximal area inscribed in a rectangle. Thus $E_A = E_I$. To prove (B), the arc length of E_0 is given by

$$L = 2 \int_0^{\pi/2} [h(t)]^{1/2} \, dt, \text{ where} \tag{2.15}$$

$$h(t) = a^2 + b^2 - \left(a^2 - b^2\right) \cos 2t. \tag{2.16}$$

The proof we give is very similar to the proof in [26] that the ellipse of maximal arc length inscribed in a square is a circle. Split up the integral in (2.15) at $\dfrac{\pi}{4}$ to obtain $\displaystyle\int_0^{\pi/4} [h(t)]^{1/2} \, dt + \int_{\pi/4}^{\pi/2} [h(t)]^{1/2} \, dt.$

In the 2nd integral, make the substitution $s = \dfrac{\pi}{2} - t$, which

yields $\displaystyle -\int_{\pi/4}^0 \left[h(\tfrac{\pi}{2} - t)\right]^{1/2} ds = -\int_{\pi/4}^0 \left[a^2 + b^2 + \left(a^2 - b^2\right) \cos 2s\right]^{1/2} ds =$

$\displaystyle\int_0^{\pi/4} \left[a^2 + b^2 + \left(a^2 - b^2\right) \cos 2t\right]^{1/2} dt;$ Thus

$$\frac{1}{2}L(u) = \int_0^{\pi/4} \left(\left[a^2 + b^2 - \left(a^2 - b^2 \right) \cos 2t \right]^{1/2} \right.$$

$$+ \left. \left[a^2 + b^2 + \left(a^2 - b^2 \right) \cos 2t \right] \right)^{1/2} dt. \tag{2.17}$$

Let $r = k^2 + l^2$ and let $w(u,t) = \sqrt{g(u)} \cos 2t$. For the rectangle Z, by Lemma 2.3 again with $d = 0$, $a^2 + b^2 = \frac{1}{2}(r + \sqrt{g(u)} + r - \sqrt{g(u)}) = r$ and $a^2 - b^2 = \frac{1}{2}(r + \sqrt{g(u)} - r + \sqrt{g(u)}) = \sqrt{g(u)}$. Thus $a^2 + b^2 \pm (a^2 - b^2) \cos 2t = r \pm \sqrt{g(u)} \cos 2t = r \pm w(u,t)$ and by (2.17),

$$\frac{1}{2}L(u) = \int_0^{\pi/4} \left[(r - w(u,t))^{1/2} + (r + w(u,t))^{1/2} \right] dt. \tag{2.18}$$

Now $g(u) \geq g(0)$ for $u \in (-1,1)$, with $g(u) = g(0) \iff u = 0$. For each $0 \leq t \leq \frac{\pi}{4}$, $\cos 2t \geq 0$, which implies that $w(u,t) \geq w(0,t)$, with equality if and only if $u = 0$. Consider the function $f(x) = (r - x)^{1/2} + (r + x)^{1/2}$, which is strictly decreasing for $0 < x < r$. Using $x = w(u,t)$, $(r - w(u,t))^{1/2} + (r + w(u,t))^{1/2} \leq (r - w(0,t))^{1/2} + (r + w(0,t))^{1/2}$, again with equality if and only if $u = 0$. Thus by (3.2), $L(u)$ attains its unique maximum on $(-1,1)$ when $u = 0$. That proves (B). ∎

Remark 2.6 *By Theorems 2.3 and 2.4, if Q is a parallelogram, then the unique ellipse of maximal area inscribed in Q equals the unique ellipse of minimal eccentricity inscribed in Q if and only if Q is a rectangle. It would be interesting to see if this holds for convex quadrilaterals in general.*

Remark 2.7 *Showing that there is a unique ellipse of maximal arc length inscribed in a general convex quadrilateral and/or characterizing such an ellipse appears to be a very nontrivial problem. Even for parallelograms it appears to be difficult. For the parallelogram $Q_{d,k,l}$, by Lemma 2.3, $a^2 + b^2 = d^2 + k^2 + l^2 + 2dlu$, which is not a constant function of u. Thus $a^2 + b^2$ does not remain constant as E_0 varies over all ellipses inscribed in $Q_{d,k,l}$. Numerical evidence suggests strongly that the ellipse of minimal eccentricity inscribed in a parallelogram, Q, is **not** the ellipse of maximal arc length inscribed in Q.*

2.5 Orthogonal Least Squares

If l is a line and P is a point in the plane, we let $d(P,l)$ denote the Euclidean distance from P to l. For the following result, see, for example, [27].

Theorem 2.5 *Let $S = \{z_1, ..., z_n\}$ be a set of n distinct points in $\mathbb{C}$, let $g = \dfrac{1}{n} \sum\limits_{j=1}^{n} z_j$ denote the centroid, and let $Z = \sum\limits_{j=1}^{n} (z_j - g)^2$.*

(a) *If $Z = 0$, then every line through g is a line of best fit for the points in S.*

(b) *If $Z \neq 0$, then the line, $\pounds$, thru g that is parallel to the vector from $(0,0)$ to $\sqrt{Z}$ is the unique line of best fit for the points in S.*

The following result establishes a connection between the foci of the ellipse of maximal area inscribed in a parallelogram, Q, and the line of best fit for the vertices of Q.

Theorem 2.6 *Let Q be a parallelogram in the xy plane with vertices A_1, A_2, A_3, and A_4. Let E_A be the unique ellipse of maximal area inscribed in Q.*

(i) *If Q is a square, then E_A is a circle, and every line through the center of Q is a line of best fit for the vertices of Q.*

(ii) *If Q is not a square, then there is a unique line, $\pounds$, which minimizes $\sum\limits_{k=1}^{4} d^2 (A_k, l)$ among all lines, l. Furthermore, the foci of E_A lie on $\pounds$.*

Remark 2.8 *Theorem 2.6 was proven in [14], but the proof given here is a bit shorter. For a different proof of Theorem 2.6 , see [1].*

Proof. Throughout, let I and J be given by (2.6). The line through the foci of an ellipse is **not** preserved in general under nonsingular affine transformations of the plane, but it is preserved under translations and rotations. By Lemma 2.4 we may assume that $Q = Q_{d,k,l}$, with $l, k, l - d > 0, d \geq 0$. Using complex notation, the vertices of $Q_{d,k,l}$ are $z_1 = -l-d-ki$, $z_2 = l-d-ki$, $z_3 = l+d+ki$, and $z_4 = -l+d+ki$, the centroid of $Q_{d,k,l}$ is $g = \sum\limits_{k=1}^{4} z_k = \hat{0}$, and Z simplifies to

$$Z = 4(I + 2dki). \tag{2.19}$$

The point of intersection of $\overleftrightarrow{D_1}$ and $\overleftrightarrow{D}_2$ is $(0,0)$ (see (2.3)). Now it is well known, and easy to show (follows from Newton's Theorem), that the center of any ellipse inscribed in a parallelogram, Q, equals the point of intersection of the diagonals of Q. As in the proof of Theorem 2.2, it follows easily that letting $u = 0$ in (2.4) yields the following equation for E_A:

$$k^2 x^2 - 2kdxy + (d^2 + l^2)y^2 - k^2 l^2 = 0. \tag{2.20}$$

Let L_A denote the line through the foci of E_A. L_A contains the center of E_A, which is $(0,0)$, and hence L_A passes through $g = (0,0)$. We want to show that

$\pounds = L_A$. Since $\pounds$ passes thru g, it suffices to prove that $\pounds$ is parallel to L_A. Note that $Q_{d,k,l}$ is a square if and only if $d = 0$ and $l = k$. To prove (i), assume that $Q_{d,k,l}$ is a square. Then $d = 0$ and $l = k$, which implies that $Z = 0$ by (2.6) and (2.19). By Theorem 2.5, then, every line through the center of $Q_{d,k,l}$ is a line of best fit for the vertices of $Q_{d,k,l}$. Letting $d = 0$ and $l = k$ in (2.20) yields the equation of E_A, $k^2x^2 + k^2y^2 - k^4 = 0$, or $x^2 + y^2 = k^2$, which is, of course, the equation of a circle. To prove (ii), we consider two cases.

Case 1: $d = 0$. Then $Q_{d,k,l}$ is a rectangle and by (2.6) and (2.19), $Z = 4(l^2 - k^2)$. We may assume that $l \neq k$ since $Q_{d,k,l}$ is not a square and hence $Z \neq 0$. If $l > k$, then $\sqrt{Z} = 2\sqrt{l^2 - k^2}$ is real, and thus the line, $\pounds$, thru g parallel to the vector from $(0,0)$ to $\sqrt{Z}$ has slope 0. Letting $d = 0$ in (2.20) yields the equation of E_A, $k^2x^2 + l^2y^2 - k^2l^2 = 0$, or

$$\frac{x^2}{l^2} + \frac{y^2}{k^2} = 1. \tag{2.21}$$

By (2.21), L_A is parallel to the x axis and passes through g, which implies that $\pounds$ is parallel to L_A and so $\pounds = L_A$. If $l < k$, then $\sqrt{Z} = \sqrt{l^2 - k^2}$ is imaginary, and thus $\pounds$ is vertical. By (2.21) again, L_A is parallel to the y axis and $\pounds$ is parallel to L_A, and again we have $\pounds = L_A$.

Case 2: $d > 0$, which implies, by (2.19), that $Z \neq 0$ since $dk \neq 0$. Let a and b denote the lengths of the semi–major and semi–minor axes, respectively, of E_A. Letting $u = 0$ in Lemma 2.3 yields $a^2 = \frac{1}{2}(J + \sqrt{I^2 + 4k^2d^2})$, $b^2 = \frac{1}{2}(J - \sqrt{I^2 + 4k^2d^2})$. Note that $a^2b^2 = \frac{1}{4}(J^2 - (I^2 + 4k^2d^2)) = k^2l^2$, which implies that $F = -a^2b^2$. We shall apply Lemma 12.6. Note that by Proposition 2.1 (with $u = 0$), $A = k^2, C = d^2 + l^2$, and $B = -2kd < 0$. Since E_A has center $= (0,0)$, $x_0 = y_0 = 0$ and the foci of E_A are $F_1 = (-\sqrt{a^2 - A}, -\sqrt{a^2 - C})$ and $F_2 = (\sqrt{a^2 - A}, \sqrt{a^2 - C})$. Hence the slope of L_A is $\dfrac{\sqrt{a^2 - C}}{\sqrt{a^2 - A}}$.

Now $\dfrac{a^2 - C}{a^2 - A} = \dfrac{\frac{1}{2}J + \frac{1}{2}\sqrt{I^2 + 4k^2d^2} - d^2 - l^2}{\frac{1}{2}J + \frac{1}{2}\sqrt{I^2 + 4k^2d^2} - k^2} = \dfrac{\sqrt{I^2 + 4k^2d^2} - I}{\sqrt{I^2 + 4k^2d^2} + I} = $

$\dfrac{4k^2d^2}{(\sqrt{I^2 + 4k^2d^2} + I)^2}$. Since $\sqrt{I^2 + 4k^2d^2} + I > \sqrt{I^2} + I = |I| + I \geq$

0, $\dfrac{\sqrt{a^2 - C}}{\sqrt{a^2 - A}} = \dfrac{2dk}{\sqrt{I^2 + 4k^2d^2} + I}$. Now it is easy to show that the follow-

ing formula holds for any complex number z with $\text{Im } z \neq 0$: $\dfrac{\text{Im } \sqrt{z}}{\text{Re } \sqrt{z}} = $

$\dfrac{|z| - \text{Re } z}{\text{Im } z}$. By (2.19) we have $\dfrac{|Z| - \text{Re } Z}{\text{Im } Z} = \dfrac{\sqrt{I^2 + 4d^2k^2} - I}{2dk} = $

$$\frac{I^2 + 4d^2k^2 - I^2}{2dk(\sqrt{I^2 + 4d^2k^2} + I)} = \frac{2dk}{\sqrt{I^2 + 4d^2k^2} + I}. \text{ Thus } \pounds \text{ is parallel to } L_A,$$

and again we have $\pounds = L_A$. That proves (ii). ■

2.6   Example

Let $d = 1$, $l = 3$, $k = 4$, so that the vertices of $Q_{d,k,l}$ are $z_1 = -4 - 4i$, $z_2 = -2 + 4i$, $z_3 = 4 + 4i$, and $z_4 = 2 - 4i$. We shall verify Theorem 2.6 and also find the unique ellipse, E_I, of minimal eccentricity inscribed in $Q_{d,k,l}$.

(i) By Proposition 2.1, the equation of any ellipse inscribed in $Q_{d,k,l}$ is $8x^2 - 4(1 + 3u)xy + (5 + 3u)y^2 + 72(u^2 - 1) = 0, -1 < u < 1$. By Theorem 2.2 and its proof and by (2.4), the unique ellipse, E_A, of maximal area inscribed in $Q_{d,k,l}$ is attained when $u = 0$ and thus E_A has equation $8x^2 - 4xy + 5y^2 - 72 = 0$. The polynomial with roots at the vertices of $Q_{d,k,l}$ is $P(z) = (z - z_1)(z - z_2)(z - z_3)(z - z_4) = z^4 + (12 - 16i)z^2 + 512 - 384i$. The foci of E_A are $F_1 = (-\sqrt{2}, -2\sqrt{2})$ and $F_2 = (\sqrt{2}, 2\sqrt{2})$. One can verify that $P'(-\sqrt{2} - 2\sqrt{2}i) = P'(\sqrt{2} + 2\sqrt{2}i) = P'(0) = 0$. $g = \frac{1}{4}(-4 - 4i - 2 + 4i + 4 + 4i + 2 - 4i) = 0$, which implies that $Z = (-4 - 4i)^2 + (-2 + 4i)^2 + (4 + 4i)^2 + (2 - 4i)^2 = -24 + 32i$. $\sqrt{Z} = \sqrt{-24 + 32i} = 2\sqrt{2} + 4\sqrt{2}i$, and the vector, $\hat{v}$, from $(0,0)$ to $\sqrt{Z}$ is given by $2\sqrt{2}\langle 1, 2\rangle$. The line, $\pounds$, thru g that is parallel to $\hat{v}$ thus has equation $y = 2x$, which implies that the line $y = 2x$ is the unique line of best fit for z_1, z_2, z_3, and z_4. Theorem 2.6 guarantees that the foci of E_A lie on $\pounds$. One can easily verify that $(\pm\sqrt{2}, \pm 2\sqrt{2})$ do lie on $\pounds$.

(ii) By Theorem 2.3 and its proof and by (2.10), the unique ellipse, E_I, of minimal eccentricity inscribed in $Q_{d,k,l}$ is attained when $u = -\frac{3}{13}$. Thus E_I has equation $169x^2 - 26xy + 91y^2 - 1440 = 0$.

2.7   Tangency Chords and Conjugate Diameters Parallel to the Diagonals

One can prove two interesting properties (perhaps mostly known) of ellipses inscribed in parallelograms which involve tangency chords and conjugate diameters.

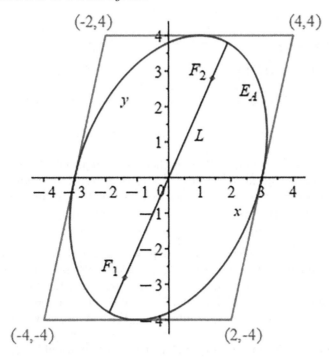

FIGURE 2.2
Ellipse of maximal area along with least squares line

(P1) Each ellipse inscribed in a parallelogram, Þ, has a pair of parallel tangency chords.

(P2) Each ellipse inscribed in a parallelogram, Þ, has a pair of conjugate diameters which are parallel (actually equal) to the diagonals of Þ.

It suffices to prove P1 and P2 above for the square with vertices $A_1 = (-1, -1)$, $A_2 = (-1, 1)$, $A_3 = (1, 1)$, and $A_4 = (1, -1)$. The details are easy and we omit them here. In addition, we prove P1 and P2 for a larger class of quadrilaterals in § 4.1 below.

Remark 2.9 *In the quadrilateral $Q_{s,t,v,w}$, if $s = v$ and $w = t - 1$, then one gets a parallelogram. One could thus work with $Q_{v,t,v,t-1}$ in all of the sections above, though we found it easier to use the quadrilateral $Q_{d,k,l}$.*

3

Area Inequality

It is well known (see [20] or [27]) that there is a unique ellipse inscribed in a given triangle, T, tangent to the sides of T at their respective midpoints. This is often called the midpoint or Steiner inellipse, and it can be characterized as the inscribed ellipse having maximum area. In addition, if E_0 is any ellipse inscribed in T, then $\text{Area}(E_0) \leq \dfrac{\pi}{3\sqrt{3}} \text{Area}(T)$, with equality if and only if E_0 is the midpoint ellipse. In this section we prove a similar inequality for ellipses inscribed in quadrilaterals (Theorem 3.1). This result was originally proven in [17], where we used a formula for the maximal area function, similar to the one given in the proof of Theorem 1.3. The proof used here is completely different and is simpler and less cumbersome. First we prove two lemmas and a proposition.

Denote the elementary symmetric functions of four variables by

$$
\begin{aligned}
\sigma_1 &= e + f + g + h, \\
\sigma_2 &= ef + eg + eh + fg + fh + gh, \\
\sigma_3 &= efg + fgh + egh + efh.
\end{aligned}
\tag{3.1}
$$

Lemma 3.1 *If* e, f, g, h *are non-negative real numbers, then* $\sigma_1^3 > 16\sigma_3$ *with equality if and only if* $e = f = g = h$.

Proof. Assuming that not all of e, f, g, h are equal, by Newton's inequalities (see [28]), $\sigma_1^2 > \dfrac{8}{3}\sigma_2$ and $\sigma_2^2 > \dfrac{9}{4}\sigma_1\sigma_3$. The latter inequality implies that $\sigma_2 > \dfrac{3}{2}\sqrt{\sigma_1\sigma_3}$ and thus the first inequality implies that $\sigma_1^2 > \dfrac{8}{3}\dfrac{3}{2}\sqrt{\sigma_1\sigma_3} = 4\sqrt{\sigma_1\sigma_3}$. Hence $\sigma_1^{3/2} > 4\sqrt{\sigma_3}$, which implies that $\sigma_1^3 > 16\sigma_3$. It is trivial that $e = f = g = h$ yields $\sigma_1^3 = 16\sigma_3$. ∎

Definition 4 *Let* E_0 *be an ellipse inscribed in a quadrilateral, Q. By the tangent lengths we mean the distances from the four vertices of Q to the points of tangency of E_0 with Q.*

In the figure below, the tangent lengths are $e, f, g,$ and h.

Recall that a tangential quadrilateral is a quadrilateral with an inscribed circle.

Lemma 3.2 *Let Q be a tangential quadrilateral. Then the tangent lengths for the inscribed circle are all equal if and only if Q is a square.*

          DOI: 10.1201/9781003474890-4

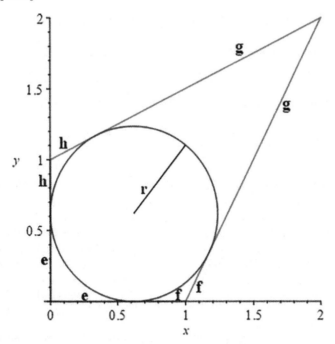

FIGURE 3.1
tangent lengths for an inscribed circle

Proof. Suppose that the tangent lengths are all equal. Using a formula for the angles of Q as functions of the tangent lengths (see Theorem 8 in [29]), each angle of Q must be 90°. By the definition of the tangent lengths, all sides of Q have equal length. Hence Q is a square. Conversely, if Q is a square, then it is easy to show that the tangent lengths for the inscribed circle are all equal. ∎

The inequality in the following proposition appears to be well-known, but this author has not found the details of a proof anywhere. For example, it was ascribed to a T. A. Ivanova in the Wikipedia article [30]. However, the reference given there for Ivanova no longer works. This inequality is also essentially given in [31], Theorem 8.1(c), but again no details of a proof. We provide a proof here, though there is probably a shorter proof out there somewhere.

Proposition 3.1 *Let Q be a tangential quadrilateral, let $A(Q)$ denote its area, let C_0 be its inscribed circle, and let r denote the radius of C_0. Then $A(Q) \geq 4r^2$, with equality if and only if Q is a square.*

Proof. Let e, f, g, h denote the tangent lengths for C_0. The following formulas can be found in several places (see the references in [29]): $r = \sqrt{\dfrac{\sigma_3}{\sigma_1}}$ and

$A(Q) = \sqrt{\sigma_1 \sigma_3}$. Suppose that Q is not a square. By Lemma 3.2, not all of the tangent lengths are equal. By Lemma 3.1, $\sigma_1^3 > 16\sigma_3 \Rightarrow \sigma_1 \sigma_3 > 16 \left(\dfrac{\sigma_3}{\sigma_1} \right)^2 \Rightarrow$ $\sqrt{\sigma_1 \sigma_3} > 4\dfrac{\sigma_3}{\sigma_1} \Rightarrow A(Q) > 4r^2$. If Q is a square, then by Lemma 3.2 again, all of the tangent lengths are equal. Then $A(Q) = \sqrt{4e(4e^3)} = 4e^2$ and $r = \sqrt{\dfrac{4e^3}{4e}} = e$, which implies that $A(Q) = 4r^2$. ∎

Theorem 3.1 : *Let E_0 be any ellipse inscribed in a convex quadrilateral, Q. Then $Area(E_0) \leq \dfrac{\pi}{4} Area(Q)$, and equality holds if and only if Q is a parallelogram and E_0 is tangent to the sides of Q at the midpoints.*

Proof. Use an affine transformation, $\maltese$ (for example, an orthogonal projection as done in [20]) to map E_0 to a circle, C_0, inscribed in the tangential quadrilateral $Q_1 = \maltese(Q)$; Since parallel lines are preserved under affine transformations, Q_1 is a square if and only if Q is a parallelogram. Since ratios of areas of ellipses are preserved under affine transformations, $\dfrac{\text{Area}(E_0)}{\text{Area}(Q)} = \dfrac{\text{Area}(C_0)}{\text{Area}(Q_1)}$; Now $\dfrac{\text{Area}(C_0)}{\text{Area}(Q_1)} = \dfrac{\pi r^2}{\text{Area}(Q_1)}$, where r is the radius of C_0. Suppose first that Q is **not** a parallelogram. Then Q_1 is **not** a square and by Proposition 3.1, $\dfrac{\pi r^2}{\text{Area}(Q_1)} < \dfrac{\pi r^2}{4r^2} = \dfrac{\pi}{4}$, which implies that $\dfrac{\text{Area}(E_0)}{\text{Area}(Q)} < \dfrac{\pi}{4}$. Now suppose that Q is a parallelogram. Then the equality in Theorem 3.1 and the tangency at the midpoints follows from Theorem 2.2 (i) and (ii). ∎

4

Midpoint Diagonal Quadrilaterals

Most of the material in this section appears in [18] with some simplified proofs and a slightly stronger result for Theorem 4.1(iii).

Definition 5 *A convex quadrilateral, Q, is called a midpoint diagonal quadrilateral if the diagonals of Q intersect at the midpoint of at least one of the diagonals of Q.*

A parallelogram, Ᵽ, is a special case of a midpoint diagonal quadrilateral since the diagonals of Ᵽ bisect one another. Equivalently, if Q is not a parallelogram, then Q is a midpoint diagonal quadrilateral if and only if the line thru the midpoints of the diagonals of Q contains one of the diagonals of Q. In light of Theorem 1.1, we also have: Q is a midpoint diagonal quadrilateral if and only if the center of any ellipse inscribed in Q lies on one of the diagonals of Q.

Remark 4.1 *After writing a paper on midpoint diagonal quadrilaterals and then this book, we became aware of two interesting articles written about them back in 2017 and in 2021. See [32] and [33]. The authors of those papers refer to them as bisect diagonal quadrilaterals. There does not appear to be any overlap between the results there and the results in [18] or in this book.*

Recall that $Q(A_1, A_2, A_3, A_4)$ denotes the quadrilateral with vertices A_1, A_2, A_3, and A_4, starting with $A_1 =$ lower left corner and going clockwise. The diagonals of $Q(A_1, A_2, A_3, A_4)$ are denoted by $D_1 = \overline{A_1 A_3}$ and $D_2 = \overline{A_2 A_4}$ and L_Q denotes the line thru the midpoints of the diagonals of Q. There are two types of midpoint diagonal quadrilaterals: Type 1, where $D_1 \subset L_Q$, which implies that the diagonals of Q intersect at the midpoint of D_2, and Type 2, where $D_2 \subset L_Q$, which implies that the diagonals of Q intersect at the midpoint of D_1. For example, consider the quadrilateral $Q = Q_{8,4,6,2}$. Then L_Q has equation $y = \frac{1}{2}x$ and $D_1 \subset L_Q$, which implies that Q is a Type 1 midpoint diagonal quadrilateral. Midpoint diagonal quadrilaterals of types 1 and 2, respectively, are illustrated below.

DOI: 10.1201/9781003474890-5

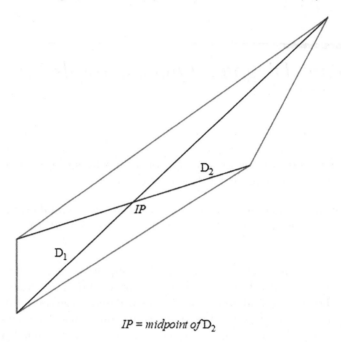

IP = midpoint of D_2

FIGURE 4.1
Type 1 midpoint diagonal quadrilateral

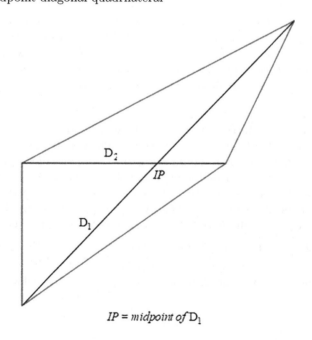

IP = midpoint of D_1

FIGURE 4.2
Type 2 midpoint diagonal quadrilateral

The following lemma shows that affine transformations preserve the class of midpoint diagonal quadrilaterals. The proof follows immediately from the properties of affine transformations.

Lemma 4.1 *Let $T : R^2 \to R^2$ be an affine transformation and let Q be a midpoint diagonal quadrilateral. Then $Q' = T(Q)$ is also a midpoint diagonal quadrilateral.*

4.1 Conjugate Diameters and Tangency Chords

Recall the following: Given a diameter, l, of an ellipse, E, there is a unique diameter, m, of E such that the midpoints of all chords parallel to l lie on m. In this case we say that l and m are **conjugate diameters** of E, or that m is a diameter of E conjugate to l. A **tangency chord** is any chord connecting two points where E is tangent to a side of the quadrilateral, Q. Recall § 2.7 in § 2:

(P1) Each ellipse inscribed in a parallelogram, Þ, has a pair of parallel tangency chords.

(P2) Each ellipse inscribed in a parallelogram, Þ, has a pair of conjugate diameters which are parallel to the diagonals of Þ.

In this section we show that P1 or P2 completely characterize the class of midpoint diagonal quadrilaterals (see Theorem 4.1 and Theorem 4.2). Thus they are a generalization of parallelograms in this sense.

Recall that $Q_{s,t}$ is the quadrilateral with vertices $A_1 = (0,0), A_2 = (0,1), A_3 = (s,t), A_4 = (1,0)$, with $s,t > 0, s + t > 1, s \neq 1$ (see (1.1)). $Q_{s,t}$ has diagonals $D_1\colon y = \dfrac{t}{s}x$ and $D_2\colon y = 1 - x$.

Lemma 4.2 *Suppose that Q is a trapezoid which is not a parallelogram. Then Q cannot be a midpoint diagonal quadrilateral.*

Proof. By affine invariance, we may assume that $Q = Q_{s,1}$ for some $0 < s \neq 1$ (see (1.1) with $t = 1$). Now let E_0 denote an ellipse inscribed in $Q_{s,1}$. The diagonal lines of $Q_{s,1}$ are $D_1\colon y = \dfrac{1}{s}x$ and $D_2\colon y = 1 - x$, and they intersect at the point $P = \left(\dfrac{s}{s+1}, \dfrac{1}{s+1} \right)$. The midpoints of the diagonal line segments are $M_1 = \left(\dfrac{s}{2}, \dfrac{1}{2} \right)$ and $M_2 = \left(\dfrac{1}{2}, \dfrac{1}{2} \right)$, respectively. $M_1 = P \iff \dfrac{s}{s+1} = \dfrac{1}{2}s$ and $\dfrac{1}{s+1} = \dfrac{1}{2}$, and each equation has the unique solution $s = 1 \notin G$. Thus $Q_{s,1}$ cannot be a type 1 midpoint diagonal quadrilateral. $M_2 = P \iff$

$\dfrac{s}{s+1} = \dfrac{1}{2}$ and $\dfrac{1}{s+1} = \dfrac{1}{2}$. Again, each equation has the unique solution $s = 1 \notin G$, and thus $Q_{s,1}$ cannot be a type 2 midpoint diagonal quadrilateral.

∎

The following lemma gives necessary and sufficient conditions for $Q_{s,t}$ to be a midpoint diagonal quadrilateral.

Lemma 4.3 *(i) $Q_{s,t}$ is a type 1 midpoint diagonal quadrilateral if and only if $s = t$.*

(ii) $Q_{s,t}$ is a type 2 midpoint diagonal quadrilateral if and only if $s+t = 2$.

Proof. By Lemma 4.2, we may assume that $t \neq 1$. The diagonal lines of $Q_{s,t}$ intersect at the point $P = \left(\dfrac{s}{s+t}, \dfrac{t}{s+t} \right)$. The midpoints of the diagonal line segments are $M_1 = \left(\dfrac{s}{2}, \dfrac{t}{2} \right)$ and $M_2 = \left(\dfrac{1}{2}, \dfrac{1}{2} \right)$, respectively. Now $M_2 = P \iff \dfrac{s}{s+t} = \dfrac{1}{2}$ and $\dfrac{t}{s+t} = \dfrac{1}{2}$, both of which hold if and only if $s = t$. That proves (i). $M_1 = P \iff \dfrac{s}{s+t} = \dfrac{1}{2}s$ and $\dfrac{t}{s+t} = \dfrac{1}{2}t$, both of which hold if and only if $s + t = 2$. That proves (ii). ∎

Theorem 4.1 *Suppose that E_0 is an ellipse inscribed in a convex quadrilateral, Q, and let S_1, S_2, S_3, and S_4 denote the sides of Q, going clockwise and starting with the leftmost side, S_1. Let $\zeta_j \in S_j, j = 1, 2, 3, 4$ denote the points of tangency of E_0 with Q. Assume that Q is not a parallelogram.*

(i) If Q is a type 1 midpoint diagonal quadrilateral, then $\overleftrightarrow{\zeta_2\zeta_3}$ and $\overleftrightarrow{\zeta_1\zeta_4}$ are parallel to D_2.

(ii) If Q is a type 2 midpoint diagonal quadrilateral, then $\overleftrightarrow{\zeta_1\zeta_2}$ and $\overleftrightarrow{\zeta_3\zeta_4}$ are parallel to D_1 .

*(iii) If Q is **not** a midpoint diagonal quadrilateral, then Q has **no** parallel tangency chords.*

Remark 4.2 *The result (iii) here is stronger than what we proved in [18], which was: If Q is **not** a midpoint diagonal quadrilateral, then neither $\overleftrightarrow{\zeta_1\zeta_2}$ nor $\overleftrightarrow{\zeta_3\zeta_4}$ are parallel to D_1, and neither $\overleftrightarrow{\zeta_2\zeta_3}$ nor $\overleftrightarrow{\zeta_1\zeta_4}$ are parallel to D_2.*

Proof. By Proposition 1.1, we may assume that Q equals the quadrilateral $Q_{s,t}$ given in (1.1), $(s,t) \in G$, where G is given in (1.2). Recall that by Proposition 1.1(i) and after simplifying, $\overleftrightarrow{\zeta_2\zeta_3}$ has slope $-\dfrac{t(2(t-1)q + s - t + 1)}{(s^2 - s + (t-1)^2 + t - 1)q + s(t - s + 1)}$, $\overleftrightarrow{\zeta_1\zeta_4}$ has slope $-\dfrac{t}{(t-s)q + s}$,

$\overleftrightarrow{\zeta_1\zeta_2}$ has slope $\dfrac{t(2(t-1)q+s)}{s((t-s)q+s)}$, and $\overleftrightarrow{\zeta_3\zeta_4}$ has slope $\dfrac{t}{(s+t-2)q+s}$. If Q is a type 1 midpoint diagonal quadrilateral, then $s=t$ by Lemma 4.3(i), which implies that the slope of $\overleftrightarrow{\zeta_2\zeta_3} = -1$ and the slope of $\overleftrightarrow{\zeta_1\zeta_4} = -1$, which proves (i). If Q is a type 2 midpoint diagonal quadrilateral, then $s+t=2$ by Lemma 4.3(ii), which implies that the slope of $\overleftrightarrow{\zeta_1\zeta_2} = \dfrac{t}{2-t}$ and the slope of $\overleftrightarrow{\zeta_3\zeta_4} = \dfrac{t}{2-t}$, which proves (ii). To prove (iii): Clearly $\overleftrightarrow{\zeta_1\zeta_2} \nparallel \overleftrightarrow{\zeta_1\zeta_3}$, $\overleftrightarrow{\zeta_1\zeta_2} \nparallel \overleftrightarrow{\zeta_1\zeta_4}, \overleftrightarrow{\zeta_3\zeta_4} \nparallel \overleftrightarrow{\zeta_1\zeta_3} \ \overleftrightarrow{\zeta_2\zeta_3} \nparallel \overleftrightarrow{\zeta_2\zeta_4}$, and $\overleftrightarrow{\zeta_1\zeta_3} \nparallel \overleftrightarrow{\zeta_2\zeta_4}$ since each of those pairs of lines must intersect. Thus we only need to prove that if Q is not a midpoint diagonal quadrilateral, then $\overleftrightarrow{\zeta_1\zeta_2} \nparallel \overleftrightarrow{\zeta_3\zeta_4}$ and $\overleftrightarrow{\zeta_2\zeta_3} \nparallel \overleftrightarrow{\zeta_1\zeta_4}$. First, $\overleftrightarrow{\zeta_1\zeta_2}$ and $\overleftrightarrow{\zeta_3\zeta_4}$ have equal slopes if and only if $(2(t-1)q+s)\left((s+t-2)q+s\right) - s((t-s)q+s) = 0 \iff 2q\,(s+t-2)\,(s+(t-1)q) = 0 \iff s+t=2$ or $q = -\dfrac{s}{t-1}$. If $t<1$, then $\dfrac{s}{1-t} > 1$ since $s+t>1$; If $t>1$, then $-\dfrac{s}{t-1} < 0$; Since $0<q<1$, q cannot equal $-\dfrac{s}{t-1}$ and thus $s+t=2$, which implies that $Q_{s,t}$ is a type 2 midpoint diagonal quadrilateral by Lemma 4.3(ii). Hence if Q is **not** a midpoint diagonal quadrilateral, then $\overleftrightarrow{\zeta_1\zeta_2} \nparallel \overleftrightarrow{\zeta_3\zeta_4}$. Second, $\overleftrightarrow{\zeta_2\zeta_3}$ and $\overleftrightarrow{\zeta_1\zeta_4}$ have equal slopes if and only if $(2(t-1)q+s-t+1)((t-s)q+s) - \left((s^2-s+(t-1)^2+t-1)q + s\,(t-s+1)\right) = 0 \iff 2\,(q-1)\,(t-s)\,(s+(t-1)q) = 0$. We already saw that $s+(t-1)q \neq 0$ and thus $s=t$, which implies that $Q_{s,t}$ is a type 1 midpoint diagonal quadrilateral by Lemma 4.3(i). Hence if Q is **not** a midpoint diagonal quadrilateral, then $\overleftrightarrow{\zeta_2\zeta_3} \nparallel \overleftrightarrow{\zeta_1\zeta_4}$. ∎

Remark 4.3 *If Q is a parallelogram, then it is easy to prove a version of Theorem 4.1(i) and (ii). Here are the details. We may assume that $Q = Q_{d,k,l}$, the parallelogram with vertices $(-l,-k), (-l,k), (l,k)$, and $(l,-k)$, where $l, k > 0, d \geq 0, d < l$. Using Proposition 2.1, it follows easily that the slope of $\overleftrightarrow{\zeta_1\zeta_2} =$ slope of $\overleftrightarrow{\zeta_3\zeta_4} = \dfrac{k}{l+d}$ and the slope of $\overleftrightarrow{\zeta_2\zeta_3} =$ slope of $\overleftrightarrow{\zeta_1\zeta_4} = \dfrac{k}{d-l}$. Since the diagonals of Q are $D_1\colon y = \dfrac{k}{l+d}x$ and $D_2\colon y = \dfrac{k}{d-l}x, \overleftrightarrow{\zeta_2\zeta_3}$ and $\overleftrightarrow{\zeta_1\zeta_4}$ are parallel to D_2 and $\overleftrightarrow{\zeta_1\zeta_2}$ and $\overleftrightarrow{\zeta_3\zeta_4}$ are parallel to D_1.*

Recall that $Q = Q(A_1, A_2, A_3, A_4)$ denotes the quadrilateral with vertices A_1, A_2, A_3, and A_4, starting with $A_1 =$ lower left corner and going clockwise. The diagonals of Q are denoted by $D_1 = \overline{A_1 A_3}$ and $D_2 = \overline{A_2 A_4}$. We also denote the lengths of the sides of Q by $a = |A_1 A_4|, b = |A_1 A_2|, c = |A_2 A_3|$, and $d = |A_3 A_4|$.

Definition 6 *If the diagonals of Q are perpendicular, then Q is called an orthodiagonal quadrilateral.*

$Q_{s,t,v,w}$ is orthogonal if and only if $(w-1)t+vs = 0$. For example, $Q_{3,4,2,1/2}$ is orthodiagonal–the diagonals are $y = \dfrac{4}{3}x$ and $y = 1 - \dfrac{3}{4}x$.

Lemma 4.4 *Suppose that* $Q = Q(A_1, A_2, A_3, A_4)$ *is a tangential quadrilateral. Then* Q *is a midpoint diagonal quadrilateral if and only if* Q *is an orthodiagonal quadrilateral.*

Proof. Since Q is tangential, there is a circle, E_0, inscribed in Q and $a + c = b + d$. Let $\zeta_j \in S_j, j = 1, 2, 3, 4$ denote the points of tangency of E_0 with Q. Define the triangles $T_1 = \triangle \zeta_4 A_1 \zeta_1$ and $T_2 = \triangle A_4 A_1 A_2$, and define the lines $L_1 = \overleftrightarrow{\zeta_1 \zeta_4}$ and $L_2 = \overleftrightarrow{\zeta_2 \zeta_3}$

(i) Suppose that Q is a midpoint diagonal quadrilateral. We prove the case when Q is type 1, the proof for type 2 being similar. Since $L_1 \parallel D_2$ by Theorem 4.1(i), T_1 and T_2 are similar triangles. Also, since E_0 is a

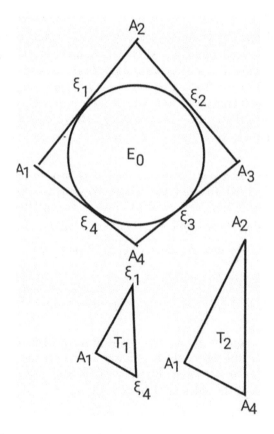

FIGURE 4.3
$E_0, T_1,$ and T_2

circle, $|A_1\zeta_4| = |A_1\zeta_1|$, which implies that T_1 is isoceles. Hence T_2 is also isoceles with $b = a$. In a similar fashion, one can show that $c = d$ using the fact that $L_2 \parallel D_2$. Thus $a^2 + c^2 = b^2 + d^2$, which implies that Q is an orthodiagonal quadrilateral.

(ii) Suppose that Q is an orthodiagonal quadrilateral. Then $a^2 + c^2 = b^2 + d^2 \Rightarrow b^2 + (a + c - b)^2 - a^2 - c^2 = 0 \Rightarrow 2(b - c)(b - a) = 0$, which implies that $a = b$ and/or $b = c$. We prove the case when $a = b$. Let $\zeta_j \in S_j, j = 1, 2, 3, 4$ denote the points of tangency of E_0 with Q. Then the triangle $T_1 = \triangle \zeta_4 A_1 \zeta_1$ is isoceles since $|A_1\zeta_4| = |A_1\zeta_1|$, and the triangle $T_2 = \triangle A_4 A_1 A_2$ is isoceles since $a = b$. Thus T_1 and T_2 are similar triangles, which implies that the line $\overleftrightarrow{\zeta_1\zeta_4}$ is parallel to $\overline{A_2 A_4} = D_2$. By Theorem 4.1(i), Q is a midpoint diagonal quadrilateral.

∎

Remark 4.4 *Though we don't use this fact, we actually proved above that if Q is tangential and a midpoint diagonal quadrilateral, then Q is a kite (two pairs of adjacent sides are equal). This does not hold in general, however. For example, $Q_{1/2,3/2}$ is a midpoint diagonal quadrilateral, but is **not** a kite.*

Theorem 4.2 *(i) Suppose that Q is a midpoint diagonal quadrilateral. Then **each** ellipse inscribed in Q has a unique pair of conjugate diameters parallel to the diagonals of Q.*

*(ii) Suppose that Q is **not** a midpoint diagonal quadrilateral. Then **no** ellipse inscribed in Q has a pair of conjugate diameters parallel to the diagonals of Q.*

Proof. Let E_0 be an ellipse inscribed in Q and let D_1 and D_2 denote the diagonals of Q. Use an affine transformation, $\mathcal{F}$, to map E_0 to a circle, E_0', inscribed in the tangential quadrilateral, $Q' = \mathcal{F}(Q)$. Let L_1 be a diameter of E_0 parallel to D_1. $\mathcal{F}$ maps L_1 to a diameter, L_1', of E_0' parallel to one of the diagonals of Q', which we call D_1'. Let D_2' be the other diagonal of Q' and let L_2' be the diameter of E_0' conjugate to L_1', which implies that $L_1' \perp L_2'$ since E_0' is a circle. By Lemma A.7, $\mathcal{F}^{-1}$ maps L_2' to L_2, a diameter of E_0 conjugate to L_1. To prove (i), suppose that Q is a midpoint diagonal quadrilateral. By Lemma 4.1, Q' is also a midpoint diagonal quadrilateral. By Lemma 4.4, Q' is an orthodiagonal quadrilateral, which implies that $D_1' \perp D_2'$. Since $L_1' \parallel D_1', L_1' \perp L_2'$, and $D_1' \perp D_2'$, L_2' must be parallel to D_2', which implies that L_2 is parallel to D_2 since $\mathcal{F}^{-1}$ is an affine transformation. That proves (i). To prove (ii), suppose that Q is **not** a midpoint diagonal quadrilateral. Since Q' is tangential, if Q' were also an orthodiagonal quadrilateral, then by Lemma 4.4, Q' would be a midpoint diagonal quadrilateral. Hence Q' cannot be an orthodiagonal quadrilateral, which implies that D_1' is not perpendicular to D_2'. Now if L_2' were parallel to D_2', then it would follow that $D_1' \perp D_2'$ since $L_1' \parallel D_1'$ and $L_1' \perp L_2'$, a contradiction. Hence $L_2' \nparallel D_2'$, and so $L_2 \nparallel D_2$ since $\mathcal{F}$ is an affine transformation. ∎

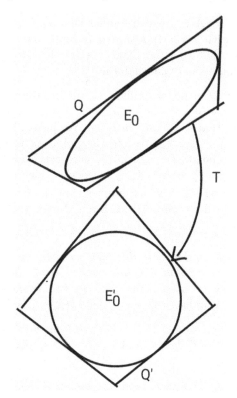

FIGURE 4.4
The map, $\mathcal{F}$, from E_0 to E_0'

4.2   Equal Conjugate Diameters and the Ellipse of Minimal Eccentricity

By Theorem 4.2(i), each ellipse inscribed in a midpoint diagonal quadrilateral, Q, has a unique pair of conjugate diameters parallel to the diagonals of Q. In particular, this holds for the unique ellipse of minimal eccentricity, E_I, inscribed in Q. However, below we prove a stronger result for E_I. Theorem 4.3(ii) was already stated and proven for parallelograms in § 2 (see Theorem 2.3), so we assume that Q is not a parallelogram.

Theorem 4.3 *Let Q be a midpoint diagonal quadrilateral which is not a parallelogram. Let E_I be the unique ellipse of minimal eccentricity in Q. Then the unique pair of conjugate diameters parallel to the diagonals of Q are* **equal** *conjugate diameters of E_I.*

Remark 4.5 *Theorem 4.3 cannot hold if Q is **not** a midpoint diagonal quadrilateral, since in that case no ellipse inscribed in Q has conjugate diameters parallel to the diagonals of Q by Theorem 4.2(ii). But Theorem 4.3 implies the following weaker result: The smallest nonnegative angle between equal conjugate diameters of E_I equals the smallest nonnegative angle between the diagonals of Q when Q is a midpoint diagonal quadrilateral. We do not know if this property of E_I can hold if Q is **not** a midpoint diagonal quadrilateral.*

Before proving Theorem 4.3, we need the following lemma, which gives necessary and sufficient conditions for $Q_{s,t,v,w}$ to be a midpoint diagonal quadrilateral.

Lemma 4.5 *(i) $Q_{s,t,v,w}$ is a type 1 midpoint diagonal quadrilateral if and only if*

$$tv = (w+1)s. \tag{4.1}$$

(ii) $Q_{s,t,v,w}$ is a type 2 midpoint diagonal quadrilateral if and only if

$$(t-2)v = (w-1)s. \tag{4.2}$$

Proof. Recall that the Newton line, $L(x)$, has equation $y = \dfrac{t}{2} + \dfrac{w+1-t}{v-s}\left(x - \dfrac{s}{2}\right)$. The diagonal line $\overleftrightarrow{D_1}$ has equation $y = \dfrac{t}{s}x$, which implies that $\overleftrightarrow{D_1} = L \iff$

$$\begin{aligned} \frac{w+1-t}{v-s} &= \frac{t}{s} \text{ and} \\ \frac{t}{2} - \frac{s}{2}\frac{w+1-t}{v-s} &= 0. \end{aligned} \tag{4.3}$$

It follows easily that (4.3) holds if and only if (4.1) holds, which proves (i). The proof of (ii) follows in a similar fashion. ∎

Remark 4.6 *Suppose that Q is a type 1 midpoint diagonal quadrilateral and let L denote the Newton Line of Q. Let CD_1 and CD_2 be the equal conjugate diameters in Theorem 4.3 parallel to the diagonals, D_1 and D_2. Let $\overleftrightarrow{CD_1}$ and $\overrightarrow{CD_2}$ denote the lines containing CD_1 and CD_2, respectively. Since $\overleftrightarrow{CD_1}$ is parallel to $\overleftrightarrow{D_1}$ and $\overleftrightarrow{D_1} = L$, $\overleftrightarrow{CD_1}$ is parallel to L. Since L and $\overleftrightarrow{CD_1}$ each pass through the center of $E_I, \overleftrightarrow{CD_1} = L$. Similarly, for type 2 midpoint diagonal quadrilaterals, $\overleftrightarrow{CD_2} = L$.*

Proof. of Theorem 4.3: We assume first that Q is a *tangential* quadrilateral. Then Q is an orthodiagonal quadrilateral by Lemma 4.4, and so the diagonals of Q are perpendicular. Also, there is a unique circle, Φ, inscribed in Q, which implies that Φ is the unique ellipse of minimal eccentricity inscribed in Q since

Φ has eccentricity 0. Since any pair of perpendicular diameters of a circle are equal conjugate diameters, in particular the unique pair which are parallel to the diagonals of Q are equal conjugate diameters of Φ, and Theorem 4.3 holds. So assume now that Q is **not** a tangential quadrilateral. It suffices to assume that $Q = Q_{s,t,v,w}$ and that (1.10), (1.12), and (1.13) hold. It follows easily that $Q_{s,t,v,w}$ is also not a tangential quadrilateral. Let E_0 be an ellipse inscribed in $Q_{s,t,v,w}$ and let a and b denote the lengths of the semi–major and semi–minor axes, respectively, of E_0. Now assume that $Q_{s,t,v,w}$ is a type 1 midpoint diagonal quadrilateral. We leave the details of the proof when $Q_{s,t,v,w}$ is a type 2 midpoint diagonal quadrilateral to the reader. Use (4.1) to substitute $\dfrac{s(w+1)}{v}$ for t in Proposition 1.3(ii). Some simplification then yields $A(r) = s(4w\,(v-s)\,r^2 - 4vwr + s\,(w+1)^2)$, $B(r) = 2sv(2\,(s-v)\,r^2 + 2vr - s\,(1+w))$, $C(r) = s^2v^2$, $D(r) = 2rs^2v\,(w-1)$, $E(r) = -2rs^2v^2$, and $F(r) = r^2s^2v^2$. Here we now refer to the proof of Theorem 1.4. Using (1.16) and (1.22), it then follows that $p(r) = -16v^2s^4(2\,(s-v)\,r + v)\alpha(r)$, where

$$\alpha(r) = 2\,(s-v)\,(v^2+w^2+1)\,r^2 + 2v\,(v^2+w^2+1)\,r - s(v^2+(w+1)^2). \quad (4.4)$$

(1.10), (1.12), and (1.13) now become $s, v > 0$ (already assumed) and $2s - v > 0$. Now $\alpha(0) = -s(v^2 + (w+1)^2) < 0$ and $\alpha(1) = s(v^2 + (w-1)^2) > 0$, which implies that α has precisely one root in J since α is a quadratic. The linear function $2\,(s-v)\,r + v$ is nonzero at $r = 0$ since $v > 0$ and at $r = 1$ since $2s - v > 0$. Thus the other factors of p are nonzero and hence p has precisely one root, $r_1 \in J$. As in the proof of Theorem 1.4, letting $r = r_1$ in Proposition 1.3(ii) yields the equation of E_I. By Theorem 4.2(i), E_I has conjugate diameters, CD_1 and CD_2, parallel to the diagonals, D_1 and D_2, of $Q_{s,t,v,w}$. Since $Q_{s,t,v,w}$ is type 1, $D_1 = L$, the line thru the midpoints of D_1 and D_2, and thus L has equation $y = \dfrac{w+1}{v}x$. It then follows that $\overleftrightarrow{CD_1} = L$ (see Remark 4.6) and hence $\overleftrightarrow{CD_1}$ has equation $y = \dfrac{w+1}{v}x$. For convenience, we let

$$\beta = (s-v)r_1 + v,$$
$$\zeta = (s-v)r_1 + s.$$

After some simplification, $\dfrac{B(r)E(r) - 2C(r)D(r)}{4A(r)C(r) - B^2(r)} = \dfrac{1}{2}\dfrac{sv}{(s-v)r+v}$. Thus by (A.4) of Lemma A.2, E_I has center (x_0, y_0), where $x_0 = \dfrac{B(r_1)E(r_1) - 2C(r_1)D(r_1)}{4A(r_1)C(r_1) - B^2(r_1)}$

$= \dfrac{1}{2}\dfrac{sv}{(s-v)r_1 + v} = \dfrac{1}{2}\dfrac{sv}{\beta}$. Since E_I has center $(x_0, L(x_0))$ by Newton's Theorem and $L\left(\dfrac{1}{2}\dfrac{sv}{\beta}\right) = \dfrac{1}{2}\dfrac{w+1}{v}\dfrac{sv}{\beta} = \dfrac{1}{2}\dfrac{s(w+1)}{\beta}$, E_I has center

$\left(\dfrac{1}{2}\dfrac{sv}{\beta}, \dfrac{1}{2}\dfrac{s(w+1)}{\beta}\right)$. Since $\overleftrightarrow{CD_2}$ passes through the center of E_I and has

the same slope as $\overleftrightarrow{D_2}$, which is $\dfrac{w-1}{v}$, $\overleftrightarrow{CD_2}$ has equation $y - \dfrac{1}{2}\dfrac{s(w+1)}{\beta} =$

$\dfrac{w-1}{v}\left(x - \dfrac{1}{2}\dfrac{sv}{\beta}\right)$, which simplifies to $y = \dfrac{w-1}{v}x + \dfrac{s}{\beta}$. Now suppose that
CD_1 intersects E_I at the two distinct points $P_j = (x_j, y_j) = (x_j, \frac{w+1}{v}x_j)$,
$j = 1, 2$, and that CD_2 intersects E_I at the two distinct points $P_j = (x_j, y_j), j = 3, 4$. Since $P_j, j = 1-4$ lies on E_I, by Proposition 1.3(ii), $A(r_1)x_j^2 + B(r_1)x_j y_j + C(r_1)y_j^2 + D(r_1)x_j + E(r_1)y_j + I(r_1) = 0, j = 1-4$. Since P_j lies on
$CD_1, j = 1, 2$, substituting $y_j = \dfrac{w+1}{v}x_j$ yields $A(r_1)x_j^2 + B(r_1)x_j^2\left(\dfrac{w+1}{v}\right) +$

$C(r_1)x_j^2\left(\dfrac{w+1}{v}\right)^2 + D(r_1)x_j + E(r_1)\left(\dfrac{w+1}{v}\right)x_j + I(r_1) = 0, j = 1, 2$, which
simplifies to

$$(A(r_1) + \left(\dfrac{w+1}{v}\right)B(r_1) + \left(\dfrac{w+1}{v}\right)^2 C(r_1))x_j^2 +$$
$$(D(r_1) + \left(\dfrac{w+1}{v}\right)E(r_1))x_j + I(r_1) = 0, j = 1, 2. \qquad (4.5)$$

If x_1 and x_2 are two distinct real roots of any quadratic, then

$$x_2 - x_1 = \dfrac{\sqrt{\tau}}{a} \Rightarrow (x_2 - x_1)^2 = \dfrac{\tau}{a^2},$$

where $\tau =$ discriminant and $a =$ leading coefficient. For the specific quadratic
in the variable x_j given in (4.5) we have

$$\tau = (D(r_1) + \left(\dfrac{w+1}{v}\right)E(r_1))^2 -$$
$$4F(r_1)(A(r_1) + \left(\dfrac{w+1}{v}\right)B(r_1) + \left(\dfrac{w+1}{v}\right)^2 C(r_1))$$

and

$$a = A(r_1) + \left(\dfrac{w+1}{v}\right)B(r_1) + \left(\dfrac{w+1}{v}\right)^2 C(r_1).$$

Now applying Proposition 1.3(ii) and simplifying yields

$$\tau = 16r_1^2 s^3 v^2 (1 - r_1)\zeta,$$
$$a = 4r_1 s \beta.$$

Hence $(x_2 - x_1)^2 = \dfrac{\tau}{a^2} = \dfrac{v^2 s(1 - r_1)\zeta}{\beta^2}$. Now $y_2 - y_1 = \dfrac{w+1}{v}(x_2 - x_1) \Rightarrow$

$(x_2 - x_1)^2 + (y_2 - y_1)^2 = (x_2 - x_1)^2 \left(1 + \left(\dfrac{w+1}{v}\right)^2\right)$, which implies that

$$(x_2 - x_1)^2 + (y_2 - y_1)^2 = \left(1 + \left(\frac{w+1}{v}\right)^2\right) \frac{v^2 s(1 - r_1)\zeta}{\beta^2}. \qquad (4.6)$$

Since P_j lies on $CD_2, j = 3, 4$, substituting $y_j = \dfrac{w-1}{v}x_j + \dfrac{s}{\beta}$ and simplifying yields

$$(A(r_1) + \frac{w-1}{v}B(r_1) + \left(\frac{w-1}{v}\right)^2 C(r_1))x_j^2 +$$

$$\left(\frac{s}{\beta}B(r_1) + \frac{2s}{\beta}\frac{w-1}{v}C(r_1) + D(r_1) + \frac{w-1}{v}E(r_1)\right)x_j + \qquad (4.7)$$

$$\left(\frac{s}{\beta}\right)^2 C(r_1) + \left(\frac{s}{\beta}\right)E(r_1) + I(r_1) = 0, j = 3, 4.$$

For the quadratic in the variable x_j given in (4.7) we have

$$\tau = \left(\frac{s}{\beta}B(r_1) + \frac{2s}{\beta}\frac{w-1}{v}C(r_1) + D(r_1) + \frac{w-1}{v}E(r_1)\right)^2$$

$$-4(A(r_1) + \frac{w-1}{v}B(r_1) +$$

$$\left(\frac{w-1}{v}\right)^2 C(r_1)) \left(\left(\frac{s}{\beta}\right)^2 C(r_1) + \left(\frac{s}{\beta}\right)E(r_1) + I(r_1)\right).$$

The leading coefficient is $a = A(r_1) + \dfrac{w-1}{v}B(r_1) + \left(\dfrac{w-1}{v}\right)^2 C(r_1)$. Applying Proposition 1.3(ii) again and simplifying yields

$$\tau = \frac{16 r_1 s^3 v^2 (r_1 - 1)^2 \zeta^2}{\beta},$$

$$a = 4s(1 - r_1)\zeta.$$

Thus $(x_4 - x_3)^2 = \dfrac{\tau}{a^2} = \dfrac{16 r_1 s^3 v^2 (r_1 - 1)^2 \zeta^2}{\beta(4s(1 - r_1)\zeta)^2} = \dfrac{r_1 s v^2}{\beta}$. Now $(y_4 - y_3)^2 =$

$\left(\dfrac{w-1}{v}\right)^2 (x_4 - x_3)^2 \Rightarrow (x_4 - x_3)^2 + (y_4 - y_3)^2 = \left(1 + \left(\dfrac{w-1}{v}\right)^2\right)(x_4 - x_3)^2$,

which implies that

$$(x_4 - x_3)^2 + (y_4 - y_3)^2 = \left(1 + \left(\frac{w-1}{v}\right)^2\right)\frac{r_1 s v^2}{\beta}. \qquad (4.8)$$

Now L_1 and L_2 are **equal** conjugate diameters if and only if $|P_1P_2| = |P_3P_4| \iff (x_2 - x_1)^2 + (y_2 - y_1)^2 = (x_4 - x_3)^2 + (y_4 - y_3)^2$. Using (4.6) and (4.8),$|P_1P_2| = |P_3P_4| \iff \left(1 + \left(\dfrac{w+1}{v}\right)^2\right)\dfrac{v^2 s(1-r_1)\zeta}{\beta^2} =$

$$\left(1 + \left(\frac{w-1}{v}\right)^2\right)\frac{r_1 s v^2}{\beta} \iff$$

$$\left(1 + \left(\frac{w+1}{v}\right)^2\right)(v^2 s(1-r_1)\zeta)\beta - \left(1 + \left(\frac{w-1}{v}\right)^2\right)r_1 s v^2 \beta^2 = 0 \iff$$

$$\left(1 + \left(\frac{w+1}{v}\right)^2\right)((1-r_1)\zeta) - \left(1 + \left(\frac{w-1}{v}\right)^2\right)r_1\beta = 0 \iff$$

$$\frac{2(s-v)(v^2 + w^2 + 1)r_1^2 + 2v(v^2 + w^2 + 1)r_1 - s(v^2 + (w+1)^2)}{v^2} = 0 \iff$$

$\alpha(r_1) = 0$, where α is given by (4.4). Since r_1 is a root of α by definition, that completes the proof of Theorem 4.3. ∎

4.3 Example

Consider the quadrilateral $Q = Q_{8,4,6,2}$, which, as noted earlier, is a type 1 midpoint diagonal quadrilateral. The diagonal lines of Q are $\overleftrightarrow{D_1}$: $y = \dfrac{1}{2}x$, which is also the equation of the Newton Line, L, and $\overleftrightarrow{D_2}$: $y = 1 + \dfrac{1}{6}x$. First, let E_0 be the ellipse inscribed in Q corresponding to $r = \dfrac{3}{7}$ in Proposition 1.3(ii). The equation of E_0 is $33x^2 - 148xy + 196y^2 + 28x - 168y = -36$. By (A.4) of Lemma A.2, E_0 has center $\left(\dfrac{7}{2}, \dfrac{7}{4}\right)$ and by Proposition 1.3(i) the points of tangency of E_0 with Q are given by $\zeta_1 = \left(0, \dfrac{3}{7}\right)$, $\zeta_2 = \left(\dfrac{32}{9}, \dfrac{7}{3}\right)$, $\zeta_3 = \left(\dfrac{62}{9}, \dfrac{26}{9}\right)$, and $\zeta_4 = \left(\dfrac{18}{7}, \dfrac{6}{7}\right)$.

- As guaranteed by Theorem 4.1(i), the slope of $\overleftrightarrow{\zeta_2\zeta_3}$ = slope of $\overleftrightarrow{\zeta_1\zeta_4}$ = $\dfrac{1}{6}$ = slope of D_2.

- Suppose that CD_1 and CD_2 are diameters of E_0 which are parallel to the diagonals D_1 and D_2, respectively, and suppose that CD_1 intersects E_0 at the two distinct points P_1 and P_2, while CD_2 intersects E_0 at

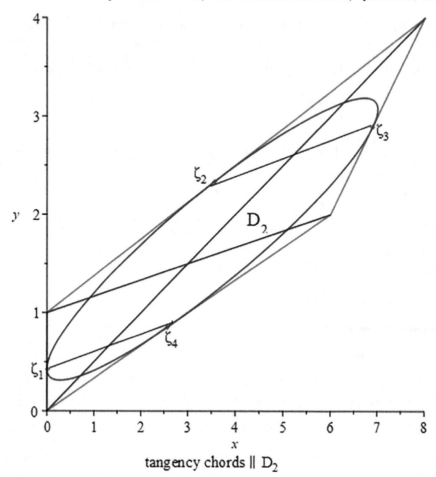

tangency chords $\parallel$ $\mathbf{D_2}$

FIGURE 4.5
Illustration of Theorem 4.1(i)

the two distinct points P_3 and P_4. The equations of $\overleftrightarrow{P_1 P_2}$ and $\overleftrightarrow{P_3 P_4}$ are thus $y - \dfrac{7}{4} = \dfrac{1}{2}\left(x - \dfrac{7}{2}\right)$ and $y - \dfrac{7}{4} = \dfrac{1}{6}\left(x - \dfrac{7}{2}\right)$, respectively. Substituting into the equation of E_0 yields $P_1 = \dfrac{1}{4}\left(14 - 2\sqrt{31}, 7 - \sqrt{31}\right)$ and $P_2 = \dfrac{1}{4}\left(14 + 2\sqrt{31}, 7 + \sqrt{31}\right)$. Similarly, $P_3 = \dfrac{1}{4}\left(14 - 6\sqrt{2}, 7 - \sqrt{2}\right)$ and $P_4 = \dfrac{1}{4}\left(14 + 6\sqrt{2}, 7 + \sqrt{2}\right)$. The family of lines which are parallel

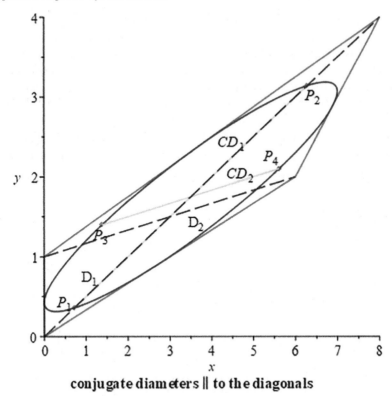

conjugate diameters ∥ to the diagonals

FIGURE 4.6
Illustration of Theorem 4.2(i)

to $\overleftrightarrow{P_1 P_2}$ have equation $y = \frac{1}{2}x + b$. Again by substituting into the equation of E_0 it is not hard to show that the chords of E_0 which are parallel to $\overleftrightarrow{P_1 P_2}$ have midpoints $\left(\frac{7}{2} - 3b, \frac{7}{4} - \frac{1}{2}b\right)$, which each lie on $\overleftrightarrow{P_3 P_4}$. Hence CD_1 and CD_2 are the pair of conjugate diameters of E_0 which are parallel to the diagonals of Q, as guaranteed by Theorem 4.2(i).

Second, let E_I be the unique ellipse of minimal eccentricity inscribed in Q. By (4.4), $\alpha(r) = 164r^2 + 492r - 360$ and the unique root of α in J is $r_1 = -\frac{3}{2} + \frac{27}{82}\sqrt{41}$. By Proposition 1.3(ii), after some simplification, the approximate equation of E_I is $70.810x^2 - 350.68xy + 552.99y^2 + 112.14x - 672.82y = -204.65$. By Theorem 4.2(i), there are conjugate diameters CD_1 and CD_2 of E_I which are parallel to the diagonals D_1 and D_2, respectively. Suppose that CD_1 intersects E_I at the two distinct points P_1 and P_2, while CD_2 intersects E_I at the two distinct points P_3 and P_4. As we did above, one can show that $P_1 \approx (1.0917, 0.54585)$, $P_2 \approx (5.5595, 2.7798)$, $P_3 \approx (0.86201, 1.2522)$, and $P_4 \approx (5.7892, 2.0734)$. That also verifies the uniqueness part of Theorem 4.2(i) as well. It then follows that $\left|\overline{P_1 P_2}\right|^2 = \left|\overline{P_3 P_4}\right|^2 \approx 24.952$. As guaranteed by Theorem 4.3, $\overline{P_1 P_2}$ and $\overline{P_3 P_4}$ are **equal** conjugate diameters of E_I which are parallel to the diagonals of Q.

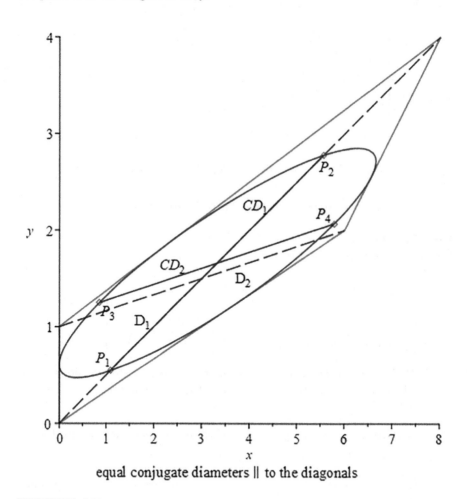

equal conjugate diameters ‖ to the diagonals

FIGURE 4.7
Illustration of Theorem 4.3

5

Tangency Points as Midpoints of Sides of Q

Among all ellipses inscribed in a triangle, T, the midpoint, or Steiner, ellipse is interesting and well–known (see [34]). It is the unique ellipse tangent to T at the midpoints of all three sides of T and is also the unique ellipse of maximal area inscribed in T. What about ellipses inscribed in quadrilaterals, Q? Not surprisingly, perhaps, there is not always a midpoint ellipse–i.e., an ellipse inscribed in Q which is tangent at the midpoints of all four sides of Q. In fact, in [1] it was shown that if there is a midpoint ellipse, then Q must be a parallelogram. That is, if Q is not a parallelogram, then there is no ellipse inscribed in Q which is tangent at the midpoint of all four sides of Q. But what about tangency at less than four sides of Q? In other words, if Q is not a parallelogram, is there an ellipse inscribed in Q which is tangent at the midpoint of two or three sides of Q (tangency at the midpoint of one side of Q is always possible)? The following results answer these questions.

Theorem 5.1 *Suppose that Q is a convex quadrilateral which is not a parallelogram. Then there is no ellipse inscribed in Q which is tangent at the midpoint of three sides of Q.*

Theorem 5.2 *Suppose that Q is a convex quadrilateral which is neither a trapezoid nor a midpoint diagonal quadrilateral. Then there is no ellipse inscribed in Q which is tangent at the midpoint of two sides of Q.*

We will prove Theorems 5.1 and 5.2 after proving two propositions below. For tangency at the midpoint of two sides of Q, we consider the cases where Q is or is not a trapezoid separately. If Q is **not** a trapezoid, then we show in Proposition 5.1(i) that there is an ellipse inscribed in Q which is tangent at the midpoint of two sides of Q if and only if Q is a midpoint diagonal quadrilateral (see §4). If Q is a trapezoid, then we show in Proposition 5.2 that there is a unique ellipse inscribed in Q which is tangent at the midpoint of two sides of Q. In addition, that ellipse is the unique ellipse of maximal area inscribed in Q. Note that in § 2 we showed that the midpoint ellipse (tangency at the midpoint of all four sides of Q) for a parallelogram also turns out to be the unique ellipse of maximal area inscribed in Q. However, for midpoint diagonal quadrilaterals, the unique ellipse of maximal area inscribed in Q need not be tangent at the midpoint of any side of Q.

DOI: 10.1201/9781003474890-6

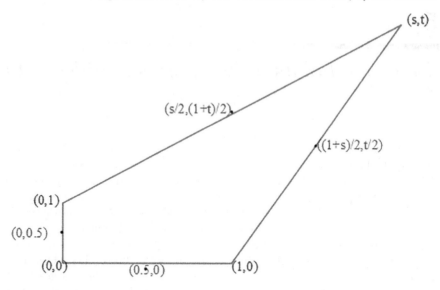

FIGURE 5.1
$Q_{s,t}$ and the midpoints of the sides

5.1 Non-Trapezoids

The result below shows that among non-trapezoids, the only quadrilaterals which have inscribed ellipses tangent at the midpoint of two sides are the midpoint diagonal quadrilaterals.

Proposition 5.1 *Let Q be a convex quadrilateral in the xy plane which is* ***not*** *a trapezoid.*

(i) There is an ellipse inscribed in Q which is tangent at the midpoint of two sides of Q if and only if Q is a midpoint diagonal quadrilateral, in which case there are two such ellipses.

(ii) There is no ellipse inscribed in Q which is tangent at the midpoint of three sides of Q.

Proof. By Lemma 4.1 and standard properties of affine transformations, we may assume that $Q = Q_{s,t}$ for some $(s,t) \in G$ (see (1.1)). The midpoints of the sides are given by $MP_1 = \left(0, \dfrac{1}{2}\right) \in S_1$, $MP_2 = \left(\dfrac{s}{2}, \dfrac{1+t}{2}\right) \in S_2$, $MP_3 = \left(\dfrac{1+s}{2}, \dfrac{t}{2}\right) \in S_3$, and $MP_4 = \left(\dfrac{1}{2}, 0\right) \in S_4$.

Let E_0 denote an ellipse inscribed in $Q_{s,t}$ and denote the corresponding points of tangency of E_0 with the sides of $Q_{s,t}$ by $\zeta_j \in S_j, j = 1, 2, 3, 4$. By Proposition 1.1(i),

$$\zeta_1 = MP_1 \iff \frac{qt}{(t-s)q+s} = \frac{1}{2}, \tag{5.1}$$

$$\zeta_2 = MP_2 \iff \frac{(1-q)s}{(t-1)(s+t)q+s} = \frac{1}{2} \text{ and} \tag{5.2}$$

$$\frac{t(s+q(t-1))}{(t-1)(s+t)q+s} = \frac{1+t}{2}. \tag{5.3}$$

$$\zeta_3 = MP_3 \iff \frac{s+q(t-1)}{(s+t-2)q+1} = \frac{1+s}{2} \text{ and} \tag{5.4}$$

$$\frac{(1-q)t}{(s+t-2)q+1} = \frac{t}{2}. \tag{5.5}$$

$$\zeta_4 = MP_4 \iff q = \frac{1}{2}. \tag{5.6}$$

Consider (5.1)–(5.6) as equations in the unknown variable q. Now (5.1) and (5.6) each have the unique solutions $q = q_1 = \dfrac{s}{s+t}$ and $q = q_4 = \dfrac{1}{2}$, respectively. The system of equations in (5.2) and (5.3) has unique solution $q = q_2 = \dfrac{s}{t(s+t-1)+s}$, and the system of equations in (5.4) and (5.5) has unique solution $q = q_3 = \dfrac{1}{s+t}$. It is trivial that $q_1, q_3, q_4 \in J = (0, 1)$. Since $(s,t) \in G$, $t(s+t-1) > 0$, which implies that $q_2 \in J$. We now check which **pairs** of midpoints of sides of $Q_{s,t}$ can be points of tangency of E_0. Note that different values of q yield distinct inscribed ellipses by the one-to-one correspondence between ellipses inscribed in $Q_{s,t}$ and points $q \in J$.

(a) S_1 and S_2: $q_1 = q_2 \iff \dfrac{s}{t(s+t-1)+s} - \dfrac{s}{s+t} = 0 \iff$
$-\dfrac{st(s+t-2)}{(ts+t^2+s-t)(s+t)} = 0 \iff s+t = 2.$

(b) S_1 and S_3: $q_1 = q_3 \iff \dfrac{1}{s+t} - \dfrac{s}{s+t} = 0 \iff -\dfrac{s-1}{s+t} = 0$, which has no solution since $(s,t) \in G$.

(c) S_2 and S_3: $q_2 = q_3 \iff \dfrac{1}{s+t} - \dfrac{s}{t(s+t-1)+s} = 0 \iff$
$\dfrac{(t-s)(s+t-1)}{(ts+t^2+s-t)(s+t)} = 0 \iff s = t.$

(d) S_1 and S_4: $q_1 = q_4 \iff \dfrac{1}{2} - \dfrac{s}{s+t} = 0 \iff -\dfrac{1}{2}\dfrac{s-t}{s+t} = 0 \iff s = t.$

(e) S_2 and S_4: $q_2 = q_4 \iff \dfrac{1}{2} - \dfrac{s}{t\,(s+t-1)+s} = 0 \iff$

$\dfrac{1}{2}\dfrac{(s+t)\,(t-1)}{ts+t^2+s-t} = 0$, which has no solution since $(s,t) \in G$ and $t \neq 1$.

(f) S_3 and S_4: $q_3 = q_4 \iff \dfrac{1}{2} - \dfrac{1}{s+t} = 0 \iff \dfrac{1}{2}\dfrac{s+t-2}{s+t} = 0 \iff$
$s+t = 2.$

That proves that there is an ellipse inscribed in $Q_{s,t}$ which is tangent at the midpoints of S_1 and S_2 or at the midpoints of S_3 and S_4 if and only if $s+t = 2$, and there is an ellipse inscribed in $Q_{s,t}$ which is tangent at the midpoints of S_2 and S_3 or at the midpoints of S_1 and S_4 if and only if $s = t$. Furthermore, if $s \neq t$ or if $s+t \neq 2$, then there is no ellipse inscribed in $Q_{s,t}$ which is tangent at the midpoint of two sides of $Q_{s,t}$. That proves (i) by Lemma 4.3. To prove (ii), to have an ellipse inscribed in $Q_{s,t}$ which is tangent at the midpoint of three sides of $Q_{s,t}$, those three sides are either S_1, S_2, and S_3 or S_1, S_2, and S_4 or S_1, S_3, and S_4 or S_2, S_3, and S_4. By (a)–(f) above, that is not possible. ∎

5.1.1   Example

Again we consider the quadrilateral $Q = Q_{8,4,6,2}$, which is a type 1 midpoint diagonal quadrilateral. The midpoints of the sides of Q are $MP_1 = \left(0, \dfrac{1}{2}\right)$, $MP_2 = \left(4, \dfrac{5}{2}\right)$, $MP_3 = (7,3)$, and $MP_4 = (3,1)$, and the points of tangency of an ellipse, E_r, inscribed in Q are $\zeta_1 = (0,r)$, $\zeta_2 = \left(48\dfrac{1-r}{4r+6}, \dfrac{-14r+24}{4r+6}\right)$, $\zeta_3 = \left(48\dfrac{8+2r}{32r+48}, \dfrac{-32r+192}{32r+48}\right)$, and $\zeta_4 = (6r, 2r)$. Proposition 5.1(i) guarantees that there are two ellipses tangent at the midpoint of two sides of Q. It then follows easily for this particular example that those two ellipses correspond to $r = \dfrac{1}{2}$ and $r = \dfrac{3}{8}$. For $r = \dfrac{1}{2}$, Q is tangent at MP_1 and at MP_4 and, by Proposition 1.3(ii), the corresponding equation is $11x^2 - 51xy + 72y^2 + 12x - 72y = -18$. For $r = \dfrac{3}{8}$, Q is tangent at MP_2 and at MP_3 and, again by Proposition 1.3(ii), the corresponding equation is $207x^2 - 909xy + 1152y^2 + 144x - 864y = -162$.

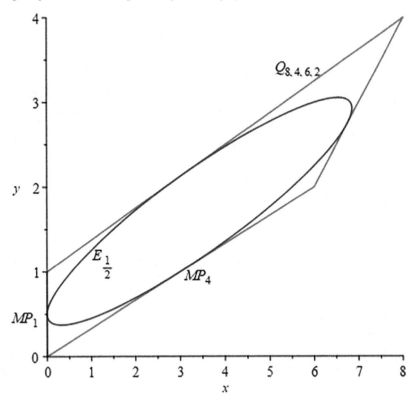

FIGURE 5.2
Ellipse tangent at MP_1 and at MP_4

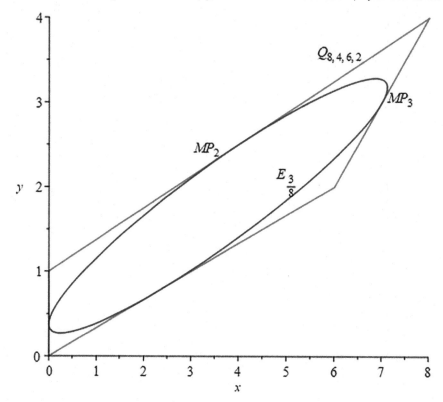

FIGURE 5.3
Ellipse tangent at MP_2 and at MP_3

5.2 Trapezoids

For trapezoids inscribed in Q we have the following result.

Proposition 5.2 *Assume that Q is a trapezoid which is not a parallelogram. Then*

(i) *There is a unique ellipse inscribed in Q which is tangent at the midpoint of two sides of Q, and that ellipse is the unique ellipse of maximal area inscribed in Q.*

(ii) *There is no ellipse inscribed in Q which is tangent at the midpoint of three sides of Q.*

Proof. Again, by affine invariance, we may assume that $Q = Q_{s,1}$ for some $0 < s \neq 1$ (see (1.1) with $t = 1$). Now let E_0 denote an ellipse inscribed in $Q_{s,1}$. Letting $MP_j \in S_j, j = 1, 2, 3, 4$ denote the corresponding midpoints of the sides and using Proposition 1.1(ii) with $t = 1$, we have

$$\zeta_1 = MP_1 \iff \frac{q}{(1-s)q + s} = \frac{1}{2}. \tag{5.7}$$

$$\zeta_2 = MP_2 \iff (1-q)s = \frac{s}{2}. \tag{5.8}$$

$$\zeta_3 = MP_3 \iff \frac{s}{(s-1)q + 1} = \frac{1+s}{2} \text{ and} \tag{5.9}$$

$$\frac{1-q}{(s-1)q + 1} = \frac{1}{2}. \tag{5.10}$$

$$\zeta_4 = MP_4 \iff q = \frac{1}{2}. \tag{5.11}$$

As before, we consider (5.7)–(5.11) as equations in the unknown variable q. Now (5.8) and (5.11) each have the unique solution $q = q_1 = \frac{1}{2} \in J$. The unique solution of (5.7) is $q = q_2 = \frac{s}{1+s} \in J$, and the unique solution of the system of equations in (5.9) and (5.10) is $q = q_3 = \frac{1}{1+s} \in J$. We now check which **pairs** of midpoints of sides of $Q_{s,1}$ can be points of tangency of E_0:

(a) $q = \frac{1}{2}$ gives tangency at the midpoints of S_2 and S_4.

(b) S_1 and S_2 or S_1 and S_4: $\frac{s}{1+s} = \frac{1}{2} \iff s = 1$.

(c) S_2 and S_3 or S_3 and S_4: $\frac{1}{1+s} = \frac{1}{2} \iff s = 1$.

(d) S_1 and S_3: $\frac{s}{1+s} = \frac{1}{1+s} \iff s = 1$.

Since we have assumed that $s \neq 1$, the only way to have an ellipse inscribed in $Q_{s,1}$ which is tangent at the midpoint of two sides of $Q_{s,1}$ is if those sides are S_2 and S_4 and $q = \frac{1}{2}$. That proves that there is a unique ellipse inscribed in $Q_{s,1}$ which is tangent at the midpoint of two sides of $Q_{s,1}$. By the proof of Theorem 1.3 with $t = 1$, the area of any ellipse inscribed in $Q_{s,1}$ is maximized when the function $f(q) = sq(1-q)$ is maximized. The latter occurs when $q = \frac{1}{2}$, which thus gives the ellipse of maximal area inscribed in $Q_{s,1}$. That proves the rest of (i). To prove (ii): If $q \neq \frac{1}{2}$, then there is no ellipse inscribed in $Q_{s,1}$ which is tangent at the midpoint of two sides of $Q_{s,1}$. To have an ellipse inscribed in $Q_{s,1}$ which is tangent at the midpoint of three sides of $Q_{s,1}$, by (a)–(d) above, those three sides must be either S_2, S_4, and S_3, or S_2, S_4, and S_1. But that is not possible by (b) and (c) above.

proof of Theorem 5.1: Since a convex quadrilateral which is not a parallelogram is either a trapezoid or not is a trapezoid, Theorem 5.1 follows immediately from Proposition 5.1(ii) and Proposition 5.2(ii). Theorem 5.2 also follows immediately from results above and we omit the details. ∎

6

Dynamics of Ellipses Inscribed in Quadrilaterals

Most of the material in this section appears in [15]. Given a quadrilateral, Q, we let int(Q) denote the interior of Q. The main result of this section is the following:

Theorem 6.1 *Let Q be a convex quadrilateral in the xy plane and let IP denote the point of intersection of the diagonals, D_1 and D_2, of Q. Let P be a point in* int$(Q) \cup Q$.

Theorem 6.2 *(i) If $P \in$ int$(Q), P \notin D_1 \cup D_2$, then there are exactly two ellipses inscribed in Q which pass through P.*

(ii) If $P \in$ int(Q) and $P \in D_1 \cup D_2$, but $P \neq IP$, then there is exactly one ellipse inscribed in Q which passes through P.

(iii) There is no ellipse inscribed in Q which passes through IP.

(iv) If $P \in Q$, but P is not one of the vertices of Q, then there is exactly one ellipse inscribed in Q which passes through P (and is thus tangent to Q at one of its sides).

We illustrate Theorem 6.1(i) and (ii) below in the examples.

Theorem 6.1 implies the following result:

Corollary 4 *If two ellipses inscribed in a convex quadrilateral intersect at a point, then that point of intersection cannot lie on either diagonal of the quadrilateral.*

Remark 6.1 *Instead of just looking at the class of ellipses inscribed in convex quadrilaterals, Q, one might see whether the results of Theorem 6.1 still hold for other families of simple closed convex curves inscribed in Q. For example, one could start with $x^n + y^n = 1$ and apply all nonsingular affine transformations to generate such a family. It seems clear geometrically (we do not have a rigorous proof) that there is no simple closed convex curve inscribed in Q which passes through the intersection point of the diagonals of Q. So Theorem 6.1(iii) would still hold for any family of simple closed convex curves inscribed in Q. However, the other parts of Theorem 6.1 would not necessarily hold since they depend on that particular family of simple closed convex curves.*

DOI: 10.1201/9781003474890-7

We shall prove Theorem 6.1 shortly. First, let $Q_{s,t}$ be the quadrilateral given in (1.1) for some $(s,t) \in G$, and define the following functions of x:

$$L_2(x) = 1 + \frac{t-1}{s}x, \tag{6.1}$$

$$L_3(x) = \frac{t}{s-1}(x-1).$$

Note that $L_2 = \overleftrightarrow{(0,1)\,(s,t)}$, $L_3 = \overleftrightarrow{(s,t)\,(1,0)}$, and $L_3(x) - L_2(x) = \frac{(s+t-1)(x-s)}{s(s-1)}$. It follows that $L_3(x) < L_2(x)$ if $s > 1$ and $1 \le x < s$. We have the following for the interior of $Q_{s,t}$, depending upon whether $s < 1$ or $s > 1$:

(i) If $s > 1$, then

$$\text{int}(Q_{s,t}) = S_1 \cup S_2, \tag{6.2}$$
$$S_1 = \{(x,y) : 0 < x \le 1, 0 < y < L_2(x)\},$$
$$S_2 = \{(x,y) : 1 \le x < s, L_3(x) < y < L_2(x)\}.$$

(ii) If $s < 1$, then

$$\text{int}(Q_{s,t}) = S_3 \cup S_4, \tag{6.3}$$
$$S_3 = \{(x,y) : 0 < x \le s, 0 < y < L_2(x)\},$$
$$S_4 = \{(x,y) : s \le x < 1, 0 < y < L_3(x)\}.$$

Let

$$
\begin{aligned}
S_5 &= \{(x,y) : 0 \le x \le s, y = L_2(x)\}, \\
S_6 &= \{(x,y) : x = 0, 0 \le y \le 1\}, \\
S_7 &= \{(x,y) : 0 \le x \le 1, y = 0\}.
\end{aligned}
$$

For $Q_{s,t}$ itself we have

(iii) If $s > 1$, then

$$Q_{s,t} = \{(x,y) : 1 \le x \le s, y = L_3(x)\} \cup S_5 \cup S_6 \cup S_7. \tag{6.4}$$

(iv) If $s < 1$, then

$$Q_{s,t} = \{(x,y) : s \le x \le 1, y = L_3(x)\} \cup S_5 \cup S_6 \cup S_7. \tag{6.5}$$

Lemma 6.1 *Define the following functions of x and y:*

$$
\begin{aligned}
g(x,y) &= (ys + (1-y)t)^2 + 4t(t-1)xy, \tag{6.6} \\
p_1(x,y) &= 2t(s-t+2)xy - 2s(s-t)y^2 - 2t(sy+tx), \\
(s,t) &\in G = \{(s,t) : s, t > 0, s+t > 1, s \ne 1\}.
\end{aligned}
$$

Then for any $(x,y) \in \text{int}(Q_{s,t})$ we have (i) $g(x,y) > 0$ and (ii) $p_1(x,y) < 0$.

Proof. Suppose that $(x, y) \in \text{int}(Q_{s,t})$ and let J denote the open interval $(0, 1)$. Since $ys + (1 - y)t$ is a linear function of y which is positive at $y = 0$ (yields $t > 0$) and at $y = 1$ (yields $s > 0$),

$$ys + (1 - y)t > 0, y \in J. \tag{6.7}$$

Hence $g(x, y) > 0$ if $t \geq 1$. Assume now that $s > 1$ and $t < 1$. By completing the square we have

$$\frac{g(x, y)}{(s - t)^2} =$$

$$\left(y + \left(\frac{t}{s - t} \right) \left(\frac{2(t - 1)x}{s - t} + 1 \right) \right)^2 +$$

$$4t^2 (1 - t) x \frac{(t - 1)x + s - t}{(s - t)^4}.$$

Using similar reasoning,

$$(t - 1)x + s - t > 0, x \in J.$$

Hence $g(x, y) > 0$ if $s > 1$ and $t < 1$. Finally, assume that $s < 1$ and $t < 1$: $\frac{\partial g(x, y)}{\partial x} = 4t(t - 1)y \neq 0$, which implies that g has no critical points in $\text{int}(Q_{s,t})$. We now check g on $Q_{s,t}$ using (6.5).

(i) $g(x, L_2(x)) = \dfrac{(s^2 - (1 - t)(s + t)x)^2}{s^2}$. Since $x \leq s$ for $(x, y) \in S_1$, $s^2 - (1 - t)(s + t)x \geq s^2 - (1 - t)(s + t)s = st(s + t - 1) > 0$, and hence $g(x, L_2(x)) > 0$.

(ii) $g(x, L_3(x)) = \dfrac{t^2((s + t - 2)x + 1 - t)^2}{(s - 1)^2}$. Since $s + t - 2 < 0$ and $s \leq x$ for $(x, y) \in S_2$, $(s + t - 2)x + 1 - t \leq (s + t - 2)s + 1 - t = (s - 1)(s + t - 1) < 0$, and hence $g(x, L_3(x)) > 0$. (iii) $g(x, 0) = t^2 > 0$. (iv) $g(0, y) = (ys + (1 - y)t)^2 > 0$ by (6.7). That proves that $g(x, y) > 0$ for any $(x, y) \in Q_{s,t}$ and thus $g(x, y) > 0$ for any $(x, y) \in \text{int}(Q_{s,t})$. Now let $L = s - t + 2$. For p_1, $\dfrac{\partial p_1(x, y)}{\partial x} = 2t((s - t + 2)y - t) =$ $0 \Rightarrow y = \dfrac{t}{s - t + 2}$. $\dfrac{\partial p_1(x, y)}{\partial y} = 4s(t - s)y + 2t((s - t + 2)x - s) =$ $0 \Rightarrow x = s\dfrac{2(s - t)y + t}{t(s - t + 2)}$. Substituting $y = \dfrac{t}{s - t + 2}$ and simplify-ing yields $x = s\dfrac{2 + 3(s - t)}{(s - t + 2)^2}$. Thus the unique critical point of p_1 is

$$cp = \left(s\frac{2 + 3(s - t)}{(s - t + 2)^2}, \frac{t}{s - t + 2} \right). \text{ Now}$$

$$p_1(cp) = -\frac{4st^2(s-t+1)}{(s-t+2)^2} < 0 \qquad (6.8)$$

since $(s,t) \in G$. We now check p_1 on $\partial(Q_{s,t})$.

Case 1: $s > 1$–then use (6.4). (a) $\{(x,y) : 1 \le x \le s, y = L_3(x)\}$:
$$p_1(x, L_3(x)) = \frac{2t^2(x-s)\,k(x)}{(s-1)^2}, \text{ where } k(x) = (s+t-2)\,x+1-t.$$
$k(1) = s-1 > 0$, $k(s) = (s-1)(s+t-1) > 0$, and $x-s \le 0$. Hence $p_1(x, L_3(x)) \le 0$.

(b) S_5: $p_1(x, L_2(x)) = \dfrac{2(x-s)\,h(x)}{s}$, where $h(x) = (t-1)(s+t)\,x+s^2$. $h(0) = s^2 > 0$, $h(s) = st(s+t-1) > 0$, and $x-s \le 0$. Hence $p_1(x, L_2(x)) \le 0$.

(c) S_6: $p_1(0,y) = -2ys((s-t)y+t) \le 0$ since $(s-t)y+t$ is positive at $y = 0$ and $y = 1$.

(d) S_7: $p_1(x,0) = -2t^2x < 0$.

Case 2: $s < 1$–then use (6.5). Since $g(1) < 0$, $g(s) < 0$, and $x-s \ge 0$, $p_1(x, L_3(x)) \le 0$. Everything else follows as in case 1 above. ∎

Since $p_1 < 0$ at its only critical point in $\mathrm{int}(Q_{s,t})$ and $p_1 \le 0$ on $Q_{s,t}$, that proves that $p_1(x,y) < 0$ for any $(x,y) \in \mathrm{int}(Q_{s,t})$.

Proof. Proof of Theorem 6.1: We shall prove the case when Q is **not** a **parallelogram** and leave the details when Q is a parallelogram for the reader. By affine invariance and Lemma A.3, it suffices to consider the quadrilateral $Q_{s,t}$ given in (1.1). Write $P = (x_0, y_0)$. For **fixed** x_0, y_0, s, t, one can rewrite the left hand side of (1.5), with $x = x_0$ and $y = y_0$, as the following polynomial in q:

$$
\begin{aligned}
p(q) &= p_2 q^2 + p_1 q + p_0, \text{ where}\\
p_2 &= g(x_0, y_0),\\
p_1 &= p_1(x_0, y_0),\\
p_0 &= (sy_0 - tx_0)^2,
\end{aligned}
$$

and where $g(x,y)$ and $p_1(x,y)$ are from Lemma 6.1. Evaluating p at the endpoints of J yields

$$
\begin{aligned}
p(0) &= (sy_0 - tx_0)^2 \ge 0, \qquad (6.9)\\
p(1) &= t^2(x_0 + y_0 - 1)^2 \ge 0.
\end{aligned}
$$

Now a simple computation yields, in simplified form, the discriminant of p:

$$p_1^2 - 4p_2p_0 = \qquad (6.10)$$
$$-16s(s-1)t^2x_0y_0(y_0 - L_2(x_0))(y_0 - L_3(x_0)),$$

where L_2 and L_3 are given by (6.1). By Lemma 6.1(i), $p_2 \neq 0$, which implies that $p'(q_0) = 0$, where $q_0 = -\dfrac{p_1}{2p_2}$. Another simple computation yields $p(q_0) = -\dfrac{p_1^2 - 4p_2 p_0}{4p_2}$, which implies, by (6.10), that

$$p(q_0) = \frac{4s(s-1)t^2 x_0 y_0 (y_0 - L_2(x_0))(y_0 - L_3(x_0))}{p_2}. \tag{6.11}$$

We now assume throughout that $s > 1$. The case $s < 1$ follows similarly and we omit the details. Suppose that $(x_0, y_0) \in \text{int}(Q_{s,t})$. By (6.11) and Lemma 6.1(i), $p(q_0) < 0$. Summarizing:

$$(x_0, y_0) \in \text{int}(Q_{s,t}) \text{ and } p'(q_0) = 0 \text{ implies that } p(q_0) < 0. \tag{6.12}$$

Note that Lemma 6.1(i) implies that p is concave up and Lemma 6.1(ii) implies that $p'(0) < 0$. Hence

$$q_0 \in J. \tag{6.13}$$

Now for given $P = (x_0, y_0) \in \text{int}(Q_{s,t})$, by Proposition 1.1(ii), the number of distinct ellipses inscribed in $Q_{s,t}$ which pass through P equals the number of distinct roots of $p(q) = 0$ in J. To prove (i), suppose that $P \notin D_1 \cup D_2$. Then $x_0 + y_0 - 1 \neq 0 \neq sy_0 - tx_0$, which implies, by (6.9), that $p(0) > 0$ and $p(1) > 0$. By (6.12) and (6.13), $p(q)$ has two distinct roots in J. To prove (ii), suppose that $P \in D_1 \cup D_2$, but $P \neq IP$. Then either $x_0 + y_0 - 1 = 0$ or $sy_0 - tx_0 = 0$, but not both, which implies, by (6.9), that $p(0) = 0$ and $p(1) > 0$, or $p(0) > 0$ and $p(1) = 0$. Again, by (6.12) and (6.13), $p(q)$ has one root in J. Finally, to prove (iii), if $P = IP$, then by (6.9), $p(q)$ vanishes at both endpoints of J, which implies that p has no roots in J. We leave the details for the proof of (iv) to the reader. ∎

Remark 6.2 *In [15] the proof of the main result (Theorem 1 in that paper) was incomplete. We neglected to prove that $q_0 \in J$, where q_0 is the critical point in the proof of Theorem 6.1 above. That is essential in proving (i) and (ii).*

6.0.1   Examples

Below we look at two examples illustrating Theorem 6.1 for specific values of s and t for the quadrilateral $Q_{s,t}$.

(1) $s = \dfrac{1}{2}, t = \dfrac{3}{4}, P = \left(\dfrac{1}{3}, \dfrac{3}{4}\right)$. Then $IP = \left(\dfrac{1}{4}, \dfrac{1}{2}\right), P \in \text{int}(Q_{s,t})$, and $P \notin D_1 \cup D_2$. By Theorem 6.1(i), there are exactly two ellipses, E_1 and E_2, inscribed in $Q_{s,t}$ which pass through P. $p_2 = \dfrac{33}{256}$, $p_1 = -\dfrac{9}{64}$, and $p_0 = \dfrac{1}{64}$, which implies that $p(q) = \dfrac{1}{256}(33q^2 - 36q + 4)$, which has roots

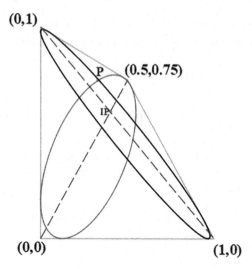

FIGURE 6.1
Illustration of Theorem 6.1(i)

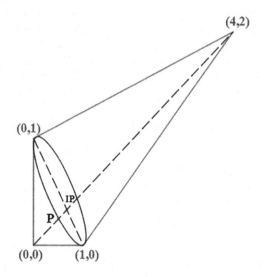

FIGURE 6.2
Illustration of Theorem 6.1(ii)

$$q_1 = \frac{1}{33}(18 - 8\sqrt{3}) \approx 0.13 \in J \text{ and } q_2 = \frac{1}{33}(18 + 8\sqrt{3}) \approx 0.97 \in J.$$

Letting $q = \dfrac{1}{33}(18 - 8\sqrt{3})$ in (1.5) yields the following approximate equation of E_1:

$562.5x^2 - 432.22xy + 282.38y^2 - 141.26x - 100.08y + 8.8685 = 0$. Letting $q = \dfrac{1}{33}(18 + 8\sqrt{3})$ in (1.5) yields the following approximate equation of E_2:

$562.5x^2 + 1085.1xy + 549.58y^2 - 1086.0x - 1073.5y + 524.19 = 0$.

(2) $s = 4$, $t = 2$, $P = \left(\dfrac{1}{2}, \dfrac{1}{4}\right)$. Then $IP = \left(\dfrac{2}{3}, \dfrac{1}{3}\right)$, $P \in \text{int}(Q_{s,t})$, $P \in D_1 \Rightarrow P \in D_1 \cup D_2$, and $P \neq IP$. By Theorem 6.1(ii), there is exactly one ellipse, E_0, inscribed in $Q_{s,t}$ which passes through P. $p_2 = \dfrac{29}{4}$, $p_1 = -7$, and $p_0 = 0$, which implies that $p(q) = \dfrac{29}{4}q^2 - 7q$, which has roots $q_1 = 0 \notin J$ and $q_2 = \dfrac{28}{29} \in J$. Letting $q = \dfrac{28}{29}$ in (1.5) yields the following equation of E_0: $841x^2 + 1452xy + 900y^2 - 1624x - 1680y + 784 = 0$.

Remark 6.3 *For the dynamics of ellispes inscribed in triangles, see [35].*

7

Algorithms for Inscribed Ellipses

Based on results from sections 1.2, 1.3, 1.5, 2.2, and 2.3, in this section we provide algorithms for finding the unique ellipse of maximal area and the unique ellipse of minimal eccentricity inscribed in a convex quadrilateral, Q. We consider the cases when Q is not a parallelogram and when Q is a parallelogram separately (one algorithm involves a trapezoid), as was done in the earlier sections. First we provide methods for linear transformations which map Q to one of the quadrilaterals considered earlier: $Q_{s,t}, Q_{s,t,v,w}$, the trapezoid $Q_{s,t,s,w}$, and the parallelogram $Q_{d,k,l}$.

7.1 Transformations

(1) Let Q be a convex quadrilateral which is **not** a parallelogram and let $Q_{s,t}$ be the convex quadrilateral with vertices $(0,0), (0,1), (s,t)$, and $(1,0)$. To find a nonsingular affine transformation, $\mathcal{F}_{s,t}$, so that $\mathcal{F}_{s,t}(Q) = Q_{s,t}$:

   (i) If Q has a pair of parallel vertical sides, first rotate counterclockwise by $90°$, yielding a quadrilateral with parallel horizontal sides.

   (ii) Write $Q = Q(A_1, A_2, A_3, A_4)$, where A_1, A_2, A_3, and A_4 are the vertices of Q, starting with the lower left corner, A_1, and going clockwise.

   (iii) Find the unique affine transformation which sends A_1, A_2, and A_4 to the points $(0,0), (0,1)$, and $(1,0)$, respectively. It then follows that $A_3 = (s,t)$ for some s,t satisfying $s,t > 0, s + t > 1, s \neq 1$.

(2) Let Q be a convex quadrilateral which is **not** a trapezoid and let $Q_{s,t,v,w}$ be the quadrilateral with vertices $(0,0), (0,1), (s,t)$, and (v,w) and where (1.10), (1.12), and (1.13) hold.

   To find an affine transformation, $\mathcal{F}_{s,t,v,w}$, which preserves the eccentricity of ellipses and so that $\mathcal{F}_{s,t,v,w}(Q) = Q_{s,t,v,w}$:

   (i) Use a translation, if necessary, to map the lower left hand corner vertex of Q to $(0,0)$.

DOI: 10.1201/9781003474890-8

(ii) Now a rotation, if necessary, yields a quadrilateral with vertices $(0,0),(0,u),(s,t)$, and (v,w), and where $s,v,u>0,t>w$.

(iii) Finally use the map $\mathbb{F}(x,y)=\left(\dfrac{1}{u}x,\dfrac{1}{u}y\right)$.

(3) Let Q be a trapezoid and let $Q_{s,t,s,w}$ be the trapezoid with vertices $(0,0),(0,1),(s,t)$, and (s,w), where $s>0$ and $t>w$.

To find an affine transformation, $\mathbb{F}_{s,t,s,w}$, which preserves the eccentricity of ellipses and so that $\mathbb{F}_{s,t,s,w}(Q)=Q_{s,t,s,w}$:

(i) Use a rotation, if necessary, to map Q to a trapezoid with parallel vertical sides.

(ii) Now a translation, if necessary, yields a quadrilateral with the lower left hand corner vertex of equal to $(0,0)$.

(iii) Finally use the map $\mathbb{F}(x,y)=\left(\dfrac{1}{u}x,\dfrac{1}{u}y\right)$.

7.2 Maximal Area and Minimal Eccentricity for Non-Parallelograms

(4) To find the unique ellipse of maximal area, E_A, inscribed in a convex quadrilateral, Q, which is **not** a parallelogram:

(i) Find a nonsingular affine transformation, $\mathbb{F}_{s,t}$, so that $\mathbb{F}_{s,t}(Q)=Q_{s,t}$ for some $(s,t)\in G$–see (1) above.

(ii) Using Lemma 1.3,

(a) If $t\neq 1$ and $t-s+2\neq 0$, then let $q_2=$
$$\frac{-(st-t+1)+\sqrt{(st-t+1)^2+s(t-1)(t-s+2)}}{(t-1)(t-s+2)}.$$

(b) If $t-s+2=0$, then let $q_2=\dfrac{1}{2}\dfrac{s}{(s-1)t+1}$.

(iii) Let $A(q),B(q),C(q),D(q),E(q)$, and $F(q)$ be the polynomials in q given by (1.6). The equation of the unique ellipse, $E_{s,t}$, of maximal area inscribed in $Q_{s,t}$ is then given by $A(q_2)x^2+B(q_2)xy+C(q_2)y^2+D(q_2)x+E(q_2)y+F(q_2)=0$.

(iv) One then has $E_A=\mathbb{F}_{s,t}^{-1}(E_{s,t})$.

(5) To find the unique ellipse of minimal eccentricity, E_I, inscribed in a convex quadrilateral, Q, which is **not** a trapezoid:

(i) Find an affine transformation, $\mathbb{F}_{s,t,v,w}$, which preserves the eccentricity of ellipses and so that $\mathbb{F}_{s,t,v,w}(Q)=Q_{s,t,v,w}$, where (1.10), (1.12), and (1.13) hold–see (2) above.

(ii) Let $A(r), B(r), C(r), D(r), E(r)$, and $F(r)$ be the polynomials in r given in (1.15), let $O(r) = A(r) + C(r), M(r) = (A(r) - C(r))^2 + (B(r))^2$, and let $p(r) = 2M(r)O'(r) - O(r)M'(r)$, a quartic polynomial. Find the unique root, r_0, of p in $J = (0,1)$.

(iii) The equation of the unique ellipse of minimal eccentricity, $E_{s,t,v,w}$, inscribed in $Q_{s,t,v,w}$ is then given by $A(r_0)x^2 + B(r_0)xy + C(r_0)y^2 + D(r_0)x + E(r_0)y + F(r_0) = 0$.

(iv) One then has $E_I = \mathfrak{T}_{s,t,v,w}^{-1}(E_{s,t,v,w})$.

(6) To find the unique ellipse of minimal eccentricity, E_I, inscribed in a trapezoid, Q, which is **not** a parallelogram:

(i) Find an affine transformation, $\mathfrak{T}_{s,t,s,w}$, which preserves the eccentricity of ellipses and so that $\mathfrak{T}_{s,t,s,w}(Q) = Q_{s,t,s,w}$ for some $s > 0$ and $t > w$–see (3) above.

(ii) Let $A(r), B(r), C(r), D(r), E(r)$, and $F(r)$ be the polynomials in r given by 1.4(ii), let $O(r) = A(r) + C(r), M(r) = (A(r)-C(r))^2+(B(r))^2$, and let $p(r) = 16s^2(t-w)q(r)$, where $q(r) = 2(t-w-1)^2 r^3 - 3(t-w-1)^2 r^2 + 2(2w-t-wt-s^2)r + s^2 + t^2$, a cubic polynomial. Find the unique root, r_0, of q in $J = (0,1)$.

(iii) The equation of the unique ellipse of minimal eccentricity, $E_{s,t,s,w}$, inscribed in $Q_{s,t,s,w}$ is then given by $A(r_0)x^2 + B(r_0)xy + C(r_0)y^2 + D(r_0)x + E(r_0)y + F(r_0) = 0$.

(iv) One then has $E_I = \mathfrak{T}_{s,t,s,w}^{-1}(E_{s,t,s,w})$

7.3 Maximal Area and Minimal Eccentricity for Parallelograms

(7) To find the unique ellipse of maximal area, E_A, inscribed in a parallelogram, Q:

(i) Let S be the square with vertices $A_1 = (-1,-1), A_2 = (-1,1), A_3 = (1,1)$, and $A_4 = (1,-1)$. Find a nonsingular affine transformation, $\mathfrak{T}$, so that $\mathfrak{T}(Q) = S$.

(ii) The ellipse of maximal area inscribed in S is the unit circle $x^2 + y^2 = 1$.

(iii) Use $\mathfrak{T}^{-1}$ to map the unit circle to E_A.

(8) To find the unique ellipse of minimal eccentricity, E_I, inscribed in a parallelogram, Q.

Let $Q_{d,k,l}$ be the parallelogram with vertices $(-l-d,-k), (-l+d,k), (l+d,k)$, and $(l-d,-k)$, where $l,k > 0, d \geq 0, d < l$.

(i) Find an isometry, $\mathcal{F}$, so that $\mathcal{F}(Q) = Q_{d,k,l}$.

(ii) The equation of the unique ellipse of minimal eccentricity, $E_{d,k,l}$, inscribed in $Q_{d,k,l}$ is then given by $k^2 x^2 - 2k(d+lu_\epsilon)xy +$
$(d^2+l^2+2dlu_\epsilon)y^2+k^2l^2(u_\epsilon^2-1) = 0$, where $u_\epsilon = -\dfrac{2dl}{d^2 + k^2 + l^2}$.

(iii) One then has $E_I = \mathcal{F}^{-1}(E_{d,k,l})$.

7.4 Dynamics

(9) Here we give an algorithm which uses the proof of Theorem 6.1.

Let Q be a convex quadrilateral in the xy plane and let $int(Q)$ denote the interior of Q. Let P be a point in $\bar{Q} = int(Q) \cup Q$. As done above for maximal area, find a nonsingular affine transformation, $\mathcal{F}_{s,t}$, so that $\mathcal{F}_{s,t}(Q) = Q_{s,t}$—see (1) above. We give the algorithm for finding all ellipses, if any, passing thru $P_{s,t} = \mathcal{F}_{s,t}(P) \in \bar{Q}_{s,t} = int(Q_{s,t}) \cup Q_{s,t}$ and inscribed in $Q_{s,t}$; Then just use $\mathcal{F}_{s,t}^{-1}$ to find all ellipses, if any, passing thru P and inscribed in Q. Let IP denote the point of intersection of the diagonals, D_1 and D_2, of $Q_{s,t}$. Write $P_{s,t} = (x_0, y_0)$ and let $p_2 = (y_0 s + (1-y_0)t)^2 + 4t(t-1)x_0 y_0$, $p_1 = 2t(s-t+2)x_0 y_0 - 2s(s-t)y_0^2 - 2t(sy_0 + tx_0)$, $p_0 = (sy_0 - tx_0)^2$, and define the quadratic in q, $p(q) = p_2 q^2 + p_1 q + p_0$; Let $A(q), B(q), C(q), D(q), E(q)$, and $F(q)$ be the polynomials in q given by (1.6).

(i) If $P_{s,t} \notin D_1 \cup D_2$, then p has two distinct roots in J, q_1 and q_2; The equation of the two ellipses, E_1 and E_2, passing thru $P_{s,t}$ and inscribed in $Q_{s,t}$ are then given by $A(q_i)x^2 + B(q_i)xy + C(q_i)y^2 + D(q_i)x + E(q_i)y + F(q_i) = 0, i = 1, 2$.

(ii) If $P_{s,t} \in D_1 \cup D_2$, but $P_{s,t} \neq IP$, then $p(q)$ has one root $q_1 \in J$; The equation of the unique ellipse, E_1, passing thru $P_{s,t}$ and inscribed in $Q_{s,t}$ is then given by $A(q_1)x^2 + B(q_1)xy + C(q_1)y^2 + D(q_1)x + E(q_1)y + F(q_1) = 0$.

(iii) If $P_{s,t} = IP$, then there is no ellipse inscribed in $Q_{s,t}$ which passes through $P_{s,t}$.

(iv) If $P_{s,t} \in Q_{s,t}$, but $P_{s,t}$ is not one of the vertices of $Q_{s,t}$, then $p(q)$ has one root $q_1 \in J$; There is exactly one ellipse inscribed in $Q_{s,t}$ which passes through $P_{s,t}$ and one proceeds as in (ii) above.

Part II

Ellipses Circumscribed about Quadrilaterals

In Part 2 we discuss ellipses **circumscribed** about convex quadrilaterals–that is, ellipses passing through the four vertices of Q. Another way of putting this is convex quadrilaterals inscribed in ellipses. Many of the results here are from [16], though we add several new results. As with inscribed ellipses in Part 1, we discuss questions of existence and uniqueness of ellipses of minimal area and ellipses of minimal eccentricity circumscribed about Q. Again the results are separated for parallelograms (Theorems 9.1 and 9.3) and for non–parallelograms (Theorems 8.1 and 8.3). Unlike for inscribed ellipses, there is a nice characterization for the ellipse of minimal eccentricity circumscribed about Q for both parallelograms (Theorem 9.2) and non–parallelograms (Theorem 8.2). The essence of the characterization is due to Steiner ([36]) and involves the unique pair of conjugate directions that belong to all ellipses which pass through the vertices of Q.

Remark 7.1 *For an interesting article about ellipses and other conics circumscribed about quadrilaterals, see [37].*

Useful for part 2 is the following well-known special quadrilateral:

Definition 7 *If there is a circle passing thru the vertices of a convex quadrilateral, Q, then Q is called a cyclic quadrilateral.*

DOI: 10.1201/9781003474890-9

8

Non-Parallelograms

8.1 Equation

Recall from § 1.3 that $Q_{s,t,v,w}$ is the quadrilateral with vertices $(0,0), (0,1), (s,t)$, and (v,w), where $s, v > 0$ and $t > w$. Going clockwise, L_1: $x = 0$, L_2: $y = 1 + \dfrac{t-1}{s}x$, L_3: $y = w + \dfrac{t-w}{s-v}(x-v)$, and L_4: $y = \dfrac{w}{v}x$ denote the lines which make up the boundary of $Q_{s,t,v,w}$. As in (1.11), we let $f_1 = v(t-1) + (1-w)s$ and $f_2 = vt - ws$. As earlier, we assume throughout this section that (1.10), (1.12), and (1.13) hold. We also let

$$f_3 = sv\,(t-w)\,, f_4 = sv\,(s-v)\,.$$

Note that $f_3 \neq 0 \neq f_4$.

Lemma 8.1 *Define the quadratic polynomials in u, $I(u) = 4f_3^2 u - \left(f_4 u + \left(t^2 v - sw^2 - f_2\right)\right)^2$ and $H(u) = f_3^2(u-1)^2 + \left(f_4 u + \left(t^2 v - sw^2 - f_2\right)\right)^2$. Then*

(i) I has two real roots

$$u = \frac{\sqrt{4\,(t-w)\,f_1 f_2 f_3 + (s-v)^2\,(sw^2 - t^2 v + f_2)^2} \pm 2\sqrt{t-w}\sqrt{f_1 f_2 f_3}}{f_4\,(s-v)}.$$

(ii) Let u_1 and u_2 be the roots in (i) with the negative or positive square root, respectively. Then $I(u) > 0$ for $u_1 < u < u_2$, $I'(0) > 0$, and $\dfrac{(t-1)w}{sv} \leq u_1$.

(iii) $H(0) > 0, H'(0) < 0$, and $H''(0) > 0$.

Proof. We prove (i)–(iii) together. First, $\dfrac{1}{2}I''(0) = -f_4^2 < 0$, which implies that I is concave down on the real line, $\Re$. Note that $H(u) + I(u) = f_3^2(u+1)^2$, which implies that $I(0) + H(0) > 0$. Since $I(0) = -\left(t^2 v - sw^2 - f_2\right)^2 \leq 0$, it follows that $H(0) > 0$. $H'(0) = -2svf_1 f_2 < 0$ and $H''(0) = 2s^2 v^2((s - v)^2 + (t - w)^2) > 0$, which proves (iii). $I'(u) = 2f_3^2(u+1) - H'(u)$, which implies that $I'(0) = 2f_3^2 - H'(0) > 0$; The discriminant of I simplifies to

DOI: 10.1201/9781003474890-10

$16s^3v^3 (t - w)^2 f_1 f_2 > 0$, which implies that I has two distinct real roots. The quadratic formula and some simplification yield u_1 and u_2 above. $I(0) \le 0$, $I'(0) > 0$, and I concave down on $\Re$ then yield the first part of (ii). $I'\left(\dfrac{(t-1)w}{sv}\right) = 2sv(t - w)(vf_1 + sf_2) > 0$, which gives $\dfrac{(t-1)w}{sv} \le u_1$ and completes the proof of (ii). $\blacksquare$

Proposition 8.1 E_0 *is an ellipse passing thru the vertices of* $Q_{s,t,v,w}$ *if and only if the general equation of* E_0 *is given by*

$$f_3 u x^2 + (sw^2 - t^2 v + f_2 - f_4 u)xy + f_3 y^2 + (f_1 tw - f_2 svu)x - f_3 y = 0,$$

$$u_1 < u < u_2, \tag{8.1}$$

where u_1 *and* u_2 *are given in Lemma 8.1. The center of* E_0 *is given by* (x_0, y_0), *where*

$$I(u)x_0 = -f_3(sw^2 - t^2 v + f_2 + 2f_1 tw - (f_4 + 2f_2 sv)u)$$
$$I(u)y_0 = (sw^2 - t^2 v + f_2)f_1 tw - \tag{8.2}$$
$$(f_1 f_4 tw + f_2 sv(sw^2 - t^2 v + f_2) - 2f_3^2)u + f_2 f_4 svu^2).$$

Proof. Suppose that E_0 is an ellipse passing thru the vertices of $Q_{s,t,v,w}$, with equation $G(x,y) = a_6 x^2 + a_5 xy + a_4 y^2 + a_3 x + a_2 y + a_1 = 0$. $G(0,0) = 0 \Rightarrow a_1 = 0$; $G(0,1) = 0 \Rightarrow a_4 + a_2 = 0$; $G(s,t) = 0$ and $G(v,w) = 0$ imply that $a_6 s^2 + a_5 st + a_4 t^2 + a_3 s - a_4 t = 0$ and $a_6 v^2 + a_5 vw + a_4 w^2 + a_3 v - a_4 w = 0$, respectively. Solving for a_5 in each of the last two equations yields, after simplifying, $a_5 = -\dfrac{a_6 s^2 + a_4 t^2 + a_3 s - a_4 t}{st}$ and $a_5 = -\dfrac{a_6 v^2 + a_4 w^2 + a_3 v - a_4 w}{vw}$. Setting the two expressions on the right hand side equal to one another and solving for a_3 yields, after simplifying, $a_3 = \dfrac{f_1 twa_4 - f_2 sva_6}{f_3}$. Substituting for a_3 in either expression for a_5 and simplifying gives $a_5 = \dfrac{(sw^2 - t^2 v + f_2) a_4 + sv (v - s) a_6}{f_3}$; Thus

$$G(x,y) = a_6 x^2 + \frac{(sw^2 - t^2 v + f_2) a_4 + sv(v - s) a_6}{f_3} xy +$$
$$a_4 y^2 + \frac{wt(s - v + f_2) a_4 - svf_2 a_6}{f_3} x - a_4 y.$$

If $a_4 = 0$, it is not hard to show that $G(x,y) = 0$ yields the union of two lines. Thus we may assume that $a_4 \ne 0$. Dividing through by a_4, letting $u = \dfrac{a_6}{a_4}$, and multiplying thru by f_3 gives $G(x,y) = Ax^2 + Bxy + Cy^2 + Dx + Ey + F$, where

$$A = f_3 u, B = sw^2 - t^2 v + f_2 - f_4 u, C = f_3, \tag{8.3}$$
$$D = f_1 tw - f_2 svu, E = -f_3, F = 0.$$

We have assumed that E_0 is an ellipse, so by Lemma 12.1 in the Appendix, $\Delta = 4AC - B^2 > 0$; Using the coefficients of $G(x, y)$ given by (8.3), one can show that $\Delta = I(u)$, where I is the quadratic from Lemma 8.1, which shows that $\Delta > 0$ for $u_1 < u < u_2$. That proves that if E_0 is an ellipse passing thru the vertices of $Q_{s,t,v,w}$, then E_0 has equation $G(x, y) = 0$ with $u_1 < u < u_2$. Now suppose that E_0 has equation $G(x, y) = Ax^2 + Bxy + Cy^2 + Dx + Ey + F = 0$ with $u_1 < u < u_2$ and where the coefficients of $G(x, y)$ are given by (8.3). It is simple to prove that $G(0, 0) = G(0, 1) = G(s, t) = G(v, w) = 0$, which implies that E_0 passes thru the vertices of $Q_{s,t,v,w}$. As already noted above, $\Delta = 4AC - B^2 > 0$ since $I(u) > 0$ for $u_1 < u < u_2$. Also, $C = f_3 > 0$ and $4AC - B^2 > 0$ implies that $A > 0$ as well. To prove that $G(x, y) = 0$ yields the equation of an ellipse, we must also show that $\delta = CD^2 + AE^2 - BDE - FA > 0$. Now $\delta = f_3 f_1 f_2(svu + (1-t)w)(svu + (1-w)t))$; note that $\dfrac{(w-1)t}{sv} - \dfrac{(t-1)w}{sv} = \dfrac{w-t}{sv} < 0 \Rightarrow \dfrac{(w-1)t}{sv} < \dfrac{(t-1)w}{sv}$. By Lemma 8.1, $\dfrac{(t-1)w}{sv} \le u_1$, so if we assume that $u_1 < u < u_2$, then $u > \dfrac{(t-1)w}{sv}$ and $u > \dfrac{(w-1)t}{sv}$. That implies that $svu + (1-t)w > 0$ and $svu + (1-w)t > 0$, which in turn implies that $\delta > 0$ since $f_3 > 0$. That proves (8.1). By Lemma A.1 again, $x_0 = \dfrac{BE - 2CD}{\Delta}$ and $y_0 = \dfrac{BD - 2AE}{\Delta}$; Using the formula for the coefficients A–F and simplifying proves (8.2). $\blacksquare$

8.2 Minimal Eccentricity

Given a convex quadrilateral, Q, for simplicity, call an ellipse which passes through the vertices of Q an ellipse of circumscription. For a pair of conjugate diameters of a given ellipse, E_0, the corresponding pair of conjugate directions are simply the slopes of the lines containing those conjugate diameters. If a conjugate diameter is vertical, then we say that its conjugate direction is ∞. In the book [36], Dörrie presents Steiner's beautiful characterization of the ellipse of circumscription which has minimal eccentricity, which he calls the most nearly circular ellipse. First, Steiner proves that there is only one pair of conjugate directions, m_1 and m_2, that belong to **all** ellipses of circumscription. Second, he proves that any ellipse whose **equal** conjugate diameters possess the directional constants, m_1 and m_2, must be an ellipse of circumscription which has minimal eccentricity. There appear to be a few gaps in Steiner's result which we fill in this section. For example, for any given convex quadrilateral, Steiner does **not** prove that there **exists** an ellipse of circumscription whose **equal** conjugate diameters possess the directional constants, m_1 and m_2, or that such an ellipse is unique. In addition, Steiner does **not** prove that there is a **unique** ellipse of circumscription which has minimal eccentricity.

That leaves open the possibility that there exists more than one ellipse of circumscription which has minimal eccentricity. First we prove (Theorem 8.1) that there is a unique ellipse of circumscription, E_O, which has minimal eccentricity. Second we prove (Theorem 8.2) that the equal conjugate diameters of E_O possess the directional constants m_1 and m_2. Finally, it is also not clear whether the quadrilateral Steiner works with includes all possible convex quadrilaterals, and in particular whether it includes parallelograms. We treat the parallelogram case below in § 9.2. In the paper [16] we worked with the same quadrilateral as Steiner worked with in [36]. Here, as for ellipses of minimal eccentricity inscribed in convex quadrilaterals in §1.3, we continue to work with the quadrilateral $Q_{s,t,v,w}$. We prove the following theorem without any discussion of conjugate directions.

Theorem 8.1 *Let Q be a convex quadrilateral in the xy plane which is **not** a parallelogram. Then there is a unique ellipse of minimal eccentricity passing thru the vertices of Q.*

Remark 8.1 *We do a similar result for parallelograms in § Ecccirpar below.*

Proof. Assume first that Q is a cyclic quadrilateral. Then by definition, there is a unique circle, Φ, passing thru the vertices of Q, which implies that Φ is the unique ellipse of minimal eccentricity circumscribed about Q since Φ has eccentricity 0. So assume now that Q is **not** a **cyclic quadrilateral**. This assumption will allow us to show that the eccentricity function (technically $\dfrac{b^2}{a^2}$) below is differentiable. Also, assume first that Q is **not** a **trapezoid**. Then it suffices to consider the case $Q = Q_{s,t,v,w}$ for some s, t, v, w, where (1.10), (1.12), and (1.13) hold. Let E_0 be an ellipse passing thru the vertices of $Q_{s,t,v,w}$ and let a and b denote the lengths of the semi–major and semi–minor axes, respectively, of E_0. As done in § 1.3, it suffices to maximize $\dfrac{b^2}{a^2}$ over all ellipses passing thru the vertices of $Q_{s,t,v,w}$. Let $H(u)$ and $I(u)$ be the quadratic polynomials from Lemma 8.1. Using (8.3), as noted above, $4AC - B^2 = I(u)$, and it also follows that $(A-C)^2 + B^2 = H(u)$. By Lemma A.1 in the Appendix, then,

$$\frac{b^2}{a^2} = \frac{I(u)}{(f_3(u+1) + \sqrt{H(u)})^2} = \frac{f_3^2(u+1)^2 - H(u)}{(f_3(u+1) + \sqrt{H(u)})^2} \text{ since } H(u) + I(u) = $$

$f_3^2(u+1)^2$. Thus we want to maximize $Z(u) = \dfrac{f_3(u+1) - \sqrt{H(u)}}{f_3(u+1) + \sqrt{H(u)}}, u_1 < u <$
u_2. Note that since Q is **not** a cyclic quadrilateral, $Q_{s,t,v,w}$ is **not** a cyclic quadrilateral and thus $(A - C)^2 + B^2 \neq 0$. That implies that $H(u) > 0$ on $\Re$ and, since $f_3(u+1) > 0$ for $0 < u_1 < u < u_2$, $Z(u)$ is differentiable on $\Re$. The numerator of $Z'(u)$ is $(f_3(u+1) + \sqrt{H(u)}) \left(f_3 - \dfrac{H'(u)}{2\sqrt{H(u)}} \right) - (f_3(u+1) - $

$\sqrt{H(u)}) \left(f_3 + \dfrac{H'(u)}{2\sqrt{H(u)}} \right) = -f_3(u+1)\dfrac{H'(u)}{\sqrt{H(u)}} + 2f_3\sqrt{H(u)}$; It follows that

$Z'(u) = 0 \iff 2H(u) = (u+1)H'(u)$ or $K(u) = 0$, where

$$K(u) = 2H(u) - (u+1)H'(u).$$

Now $K'(u) = H'(u) - (u+1)H''(u)$ and $K''(u) = H''(u) - (u+1)H'''(u) - H''(u) = -(u+1)H'''(u) = 0$, which implies that K is a linear function of u; Also, $K(0) = 2H(0) - H'(0) > 0$ and $K'(0) = H'(0) - H''(0) < 0$ by Lemma 8.1(iii). Thus $K(u) = 0$ has the unique solution $u = u_0 = -\dfrac{K(0)}{K'(0)} > 0$. A simple computation gives

$$u_0 = \frac{\left(sw^2 - t^2v + f_2\right)^2 + f_3^2 + svf_1f_2}{f_3^2 + f_4^2 + svf_1f_2}, \qquad (8.4)$$

which is the unique critical point of $Z(u)$. We now show that $Z(u_0)$ is the global maximum of $Z(u)$ on $[u_1, u_2]$. First, $K(u_0) = 0 \Rightarrow 2H(u_0) - (u_0+1)H'(u_0) = 0$; $I'(u_0) = 2f_3^2(u_0+1) - H'(u_0) = 2f_3^2(u_0+1) - \dfrac{2H(u_0)}{u_0+1} = 2f_3^2(u_0+1) - \dfrac{2(f_3^2(u_0+1)^2 - I(u_0))}{u_0+1} = \dfrac{2I(u_0)}{u_0+1}$; One can also write $I'(u_0) = \dfrac{8f_1f_2f_3^2}{sv((s-v)^2 + (t-w)^2) + f_1f_2} > 0$; Since $u_0 > 0$ as well, that implies that $I(u_0) = \dfrac{1}{2}(u_0+1)I'(u_0) > 0$; By Lemma 8.1, it follows easily that $u_1 < u_0 < u_2$; Since $Z(u) = \dfrac{I(u)}{\left(f_3(u+1) + \sqrt{H(u)}\right)^2}$ and $I(u) > 0$ for $u_1 < u < u_2$, it follows that $Z(u) > 0, u_1 < u < u_2$. In addition, $I(u_1) = I(u_2) = 0 \Rightarrow Z(u_1) = Z(u_2) = 0$; Since u_0 is the only critical point of $Z(u)$, $Z(u_0)$ is the global maximum of $Z(u)$ on $[u_1, u_2]$. That proves Theorem 8.1 when Q is not a trapezoid. If Q is a trapezoid, then it suffices to let $Q = Q_{s,t,s,w}$, the trapezoid used in § 1.5 for inscribed ellipses. Let E_0 be an ellipse circumscribed about $Q_{s,t,s,w}$ and let a and b denote the lengths of the semi–major and semi–minor axes, respectively, of E_0. By (8.1) with $v = s$, the equation of E_0 is given by

$sux^2 - (t+w-1)xy + sy^2 - (s^2u - tw)x - sy = 0, u > \dfrac{(t+w-1)^2}{4s^2}$. Proceed-

ing as above, one can show that $\dfrac{b^2}{a^2} = Z(u) = \dfrac{s^2(t-w)(u+1) - \sqrt{H(u)}}{s^2(t-w)(u+1) + \sqrt{H(u)}}$,

where $H(u) = s^2(t-w)^2(s^2u^2 - 2s^2u + (t+w-1)^2 + s^2)$; It then follows that the unique maximum of $Z(u)$ on $\left(\dfrac{(t+w-1)^2}{4s^2}, \infty\right)$ occurs for

$u = u_0 = \dfrac{(t+w-1)^2 + 2s^2}{2s^2}$. ∎

Proposition 8.2 *Let Q be a convex quadrilateral in the xy plane which is* **not** *a parallelogram. Then there is only one pair of conjugate directions, m_1*

and m_2, belonging to all of the ellipses which pass thru the vertices of Q. In addition, each ellipse which passes thru the vertices of Q possesses those conjugate direction pairs.

Remark 8.2 *Again, we do a similar result for parallelograms in § Ecccirpar below.*

Proof. Assume first that Q is not a trapezoid. Then we may assume that $Q = Q_{s,t,v,w}$ for some s, t, v, w, where (1.10), (1.12), and (1.13) hold. We follow a somewhat similar approach to that of Steiner in the Dorrie book mentioned above ([36]), but for the quadrilateral $Q_{s,t,v,w}$. We use the following well-known fact: Suppose that DI_1 and DI_2 are diameters of an ellipse, E_0; Then DI_1 and DI_2 are *conjugate* diameters if and only if the tangent line to E_0 at an endpoint of one diameter is parallel to the other diameter. Suppose that E_0 is an ellipse which passes thru the vertices of $Q_{s,t,v,w}$. Let DI_1 be a diameter of E_0 with equation $y = mx + b_1$ and let DI_2 be a diameter of E_0 with equation $y = m'x + b_2$; Suppose that DI_1 intersects E_0 at the two distinct points $P_j = (x_j, y_j), j = 1, 2$ and let T_1 be the tangent line to E_0 at P_1. Differentiating both sides of (8.1) with respect to x and substituting $x = x_1, y = y_1 = mx_1 + b_1$

yields slope of $T_1 = -\dfrac{2Ax_1 + By_1 + D}{Bx_1 + 2Cy_1 + E} = -\dfrac{2Ax_1 + B(mx_1 + b_1) + D}{Bx_1 + 2C(mx_1 + b_1) + E}$, where

A–E are given by (8.3). By Lemma A.1, the center of E_0 is (x_0, y_0), where $x_0 = \dfrac{BE - 2CD}{4AC - B^2}, y_0 = \dfrac{BD - 2AE}{4AC - B^2}$. Since DI_1 passes thru (x_0, y_0), $mx_0 + b_1 = y_0$,

which implies that $b_1 = \dfrac{BD - 2AE}{4AC - B^2} - \dfrac{BE - 2CD}{4AC - B^2}m$, and after simplifying,

that gives the slope of $T_1 = -\dfrac{2A + mB}{B + 2mC}$. It follows that DI_1 and DI_2 are

conjugate diameters if and only if $m' = -\dfrac{2A + mB}{B + 2mC}$. Now substitute for A–C using (8.3) to obtain

$$m' = -\frac{(sv\,(v-s)\,m + 2f_3)u + m(sw^2 - t^2v + f_2)}{sv\,(v-s)\,u + sw^2 - t^2v + f_2 + 2f_3m}. \tag{8.5}$$

Thus for each value of $u, u_1 < u < u_2$, (8.5) yields a pair of conjugate directions, m and m', for an ellipse which passes thru the vertices of $Q_{s,t,v,w}$. For now ignore the fact that certain values of u do not yield an ellipse. We want to find values of m and m' so that (8.5) holds for **all** values of $u, u_1 < u < u_2$. First let $u = 0$ to get

$$m' = -\frac{m(sw^2 - t^2v + f_2)}{sw^2 - t^2v + f_2 + 2f_3m}. \tag{8.6}$$

Second, let $u \to \infty$ to get

$$m' = -\frac{sv\,(v-s)\,m + 2f_3}{sv\,(v-s)}. \tag{8.7}$$

(8.6) and (8.7) imply that

$$\frac{m(sw^2 - t^2v + f_2)}{sw^2 - t^2v + f_2 + 2f_3m} = \frac{sv\,(v-s)\,m + 2f_3}{sv\,(v-s)}. \tag{8.8}$$

Solving (8.8) yields the quadratic equation in m, $sv\,(v-s)\,m^2 + 2f_3 m + sw^2 - t^2v + f_2 = 0$, with roots

$$m_1 = \frac{-f_3 - \sqrt{svf_1f_2}}{sv\,(v-s)}, m_2 = \frac{-f_3 + \sqrt{svf_1f_2}}{sv\,(v-s)}. \tag{8.9}$$

Now substitute $m = m_1$ into the right hand side of (8.5). A bit of simplification yields $m' = m_2$ for *any* value of u, which implies that (8.5) holds with $m = m_1$ and $m' = m_2$ for all values of u, and in particular for $u_1 < u < u_2$. Thus the only pair of conjugate directions belonging to all of the ellipses which pass thru the vertices of $Q_{s,t,v,w}$ are m_1 and m_2. In addition, each ellipse which passes thru the vertices of $Q_{s,t,v,w}$ possesses the conjugate direction pairs m_1 and m_2. If Q is a trapezoid, then it again suffices to let $Q = Q_{s,t,s,w}$. One proceeds as above, except now we want the slope of $T_1 = -\dfrac{2A + mB}{B + 2mC} = \pm\infty$; Hence $B + 2mC = 0$, which, upon letting $v = s$, implies that $m_2 = \dfrac{1}{2}\dfrac{t + w - 1}{s}$. The conjugate directions belonging to all of the ellipses which pass thru the vertices of $Q_{s,t,s,w}$ are then $m_1 = \infty$ and $m_2 = \dfrac{1}{2}\dfrac{t + w - 1}{s}$. Note that this can also be obtained by letting v approach s in (8.9) (the proof that this limiting argument works requires a few more details). ∎

Theorem 8.2 *Let Q be a convex quadrilateral in the xy plane which is **not** a parallelogram and is **not** a cyclic quadrilateral. Let m_1 and m_2 be the unique pair of conjugate directions belonging to all of the ellipses of circumscription and let E_0 be any ellipse of circumscription. Then the equal conjugate diameters of E_0 possess the directional constants m_1 and m_2 if and only if E_0 is the unique ellipse of circumscription of minimal eccentricity.*

Remark 8.3 *As noted above, if Q is a cyclic quadrilateral, then the unique ellipse of circumscription of minimal eccentricity is the unique circle, Φ, passing thru the vertices of Q. But every pair of conjugate diameters of Φ are equal, so Theorem 8.2 would not hold for cyclic quadrilaterals.*

Proof. Assume that Q is not a trapezoid–we omit the details for the trapezoid case. As above, we may assume that $Q = Q_{s,t,v,w}$. Let m_1 and m_2 be the unique pair of conjugate directions given in (8.9) and belonging to all of the ellipses of circumscription. Let E_0 be an ellipse passing thru the vertices of $Q_{s,t,v,w}$ and let CD_1 and CD_2 be the conjugate diameters of E_0 which possess the directional constants m_1 and m_2, respectively. Suppose that CD_1 and CD_2 have equations $y = m_1x + b_1$ and $y = m_2x + b_2$, respectively, and that CD_1 intersects E_0 at the two distinct points $P_j = (x_j, y_j), j = 1, 2$ and that

CD_2 intersects E_0 at the two distinct points $P_j = (x_j, y_j), j = 3, 4$; CD_1 and CD_2 are **equal** conjugate diameters if and only if $|P_1P_2|^2 = |P_3P_4|^2 \iff$ $(x_2 - x_1)^2 + (y_2 - y_1)^2 = (x_4 - x_3)^2 + (y_4 - y_3)^2 \iff$

$$(x_2 - x_1)^2 \left(1 + m_1^2\right) = (x_4 - x_3)^2 \left(1 + m_2^2\right) \qquad (8.10)$$

since $y_1 = m_1x_1 + b_1$, $y_2 = m_1x_2 + b_1$, $y_3 = m_2x_3 + b_2$, and $y_4 = m_2x_4 + b_2$. We shall prove that (8.10) holds if and only if $u = u_0$, where u_0 is given in (8.4). Let $G(x, y) = Ax^2 + Bxy + Cy^2 + Dx + Ey + F$, where A–F are given by (8.3). Now for any coefficients A–F, it follows immediately that

$$G(mx + b) = (A + mB + m^2C)x^2 + (Bb + 2mbC + D + Em)x + Cb^2 + Eb. \quad (8.11)$$

Now $G(x_j, m_1x_j + b_1) = 0, j = 1, 2$ since $P_j, j = 1, 2$ lies on E_O and $G(x_j, m_2x_j + b_2) = 0, j = 3, 4$ since $P_j, j = 3, 4$ lies on E_O. (8.11) then yields $(A + m_1B + m_1^2C)x_j^2 + (Bb_1 + 2m_1b_1C + D + Em_1)x_j + Cb_1^2 + Eb_1 = 0, j = 1, 2$ and $(A + m_2B + m_2^2C)x_j^2 + (Bb_2 + 2m_2b_2C + D + Em_2)x_j + Cb_2^2 + Eb_2 = 0, j = 3, 4$. Using the quadratic formula and simplifying gives

$$(x_2 - x_1)^2 = \qquad (8.12)$$
$$\frac{((B + 2m_1C)b_1 + D + Em_1)^2 - 4(A + m_1B + m_1^2C)(Cb_1^2 + Eb_1)}{(A + m_1B + m_1^2C)^2}.$$

and

$$(x_4 - x_3)^2 = \qquad (8.13)$$
$$\frac{((B + 2m_2C)b_2 + D + Em_2)^2 - 4(A + m_2B + m_2^2C)(Cb_2^2 + Eb_2)}{(A + m_2B + m_2^2C)^2}.$$

As done in the proof of Proposition 8.2,

$$b_1 = \frac{BD - 2AE}{4AC - B^2} - \frac{BE - 2CD}{4AC - B^2}m_1 \qquad (8.14)$$
$$b_2 = \frac{BD - 2AE}{4AC - B^2} - \frac{BE - 2CD}{4AC - B^2}m_2. \qquad (8.15)$$

Use (8.14) and (8.15) to substitute into the right hand sides of (8.12) and (8.13), but **without** actually substituting for A–F using (8.3) or for m_1 and m_2 using (8.9), which makes the computations much simpler. One can show that

$$(x_2 - x_1)^2 = \frac{4\left(AE^2 + D^2C - BDE\right)}{(A + m_1B + m_1^2C)(4AC - B^2)},$$
$$(x_4 - x_3)^2 = \frac{4\left(AE^2 + D^2C - BDE\right)}{(A + m_2B + m_2^2C)(4AC - B^2)}.$$

That is, these are algebraic identities. Thus (8.10) holds if and only if

$$\frac{4\left(AE^2 + D^2C - BDE\right)}{\left(A + m_1B + m_1^2C\right)\left(4AC - B^2\right)}\left(1 + m_1^2\right)$$

$$= \frac{4\left(AE^2 + D^2C - BDE\right)}{\left(A + m_2B + m_2^2C\right)\left(4AC - B^2\right)}\left(1 + m_2^2\right),$$

which simplifies to $\dfrac{1 + m_1^2}{A + m_1B + m_1^2C} = \dfrac{1 + m_2^2}{A + m_2B + m_2^2C}$ or

$$(1 + m_1^2)(A + m_2B + m_2^2C) - (1 + m_2^2)(A + m_1B + m_1^2C) = 0. \qquad (8.16)$$

The left hand side of (8.16) factors as

$$(m_1 - m_2)\,(m_1 + m_2)\,A + (m_1 - m_2)\,(m_2m_1 - 1)B - (m_1 - m_2)\,(m_1 + m_2)\,C.$$

Thus upon dividing thru by $m_1 - m_2$, (8.10) holds if and only if

$$(m_1 + m_2)\,A + (m_2m_1 - 1)B - (m_1 + m_2)\,C = 0. \qquad (8.17)$$

At this point we now substitute for A–C using (8.3) and substitute for m_1 and m_2 using (8.9). Then it follows easily that (8.17) holds if and only if
$(-f_3^2 - svf_1f_2 - (sv\,(v - s))^2)sv\,(v - s)\,u + (f_3^2 - svf_1f_2 - (sv\,(v - s))^2)(sw^2 - t^2v + f_2) + 2f_3^2 sv\,(v - s) = 0 \iff$

$$u = \frac{(f_3^2 - svf_1f_2 - (sv\,(v - s))^2)(sw^2 - t^2v + f_2) + 2f_3^2 sv\,(v - s)}{(f_3^2 + svf_1f_2 + (sv\,(v - s))^2)sv\,(v - s)}, \text{ which}$$

equals u_0 in (8.4). ∎

8.3 Minimal Area

Theorem 8.3 *Let Q be a convex quadrilateral in the xy plane which is **not** a parallelogram. Then there is a unique ellipse of minimal area passing thru the vertices of Q.*

Proof. As done for inscribed ellipses (see § 1.2), it suffices to consider the quadrilateral $Q_{s,t}$, which can be obtained from $Q_{s,t,v,w}$ by letting $v = 1$ and $w = 0$. All of the formulas done above in § 8.1 still hold with those specific values of v and of w. In particular, substituting $v = 1$ and $w = 0$ into (8.3) yields

$$\Delta = I(u) = -s^2\,(s - 1)^2\,u^2 + 2st\,(t + st + s - 1)\,u - t^2\,(t - 1)^2, \qquad (8.18)$$

where I is the quadratic from Lemma 8.1. Now $I(u) > 0$ for $u_1 < u < u_2$, where u_1 and u_2 equal $t\dfrac{st + s + t - 1 \pm 2\sqrt{st}\sqrt{s + t - 1}}{s\,(s - 1)^2}$. In addition,

$\delta = CD^2 + AE^2 - BDE - F\Delta = \dfrac{s+t-1}{st}u(su+t)$. Note that by Lemma 8.1(ii), $0 \leq u_1$. Let E_0 denote an ellipse passing thru the vertices of $Q_{s,t}$ and let a and b denote the lengths of the semi–major and semi–minor axes, respectively, of E_0. By Lemma A.2 in the Appendix, $a^2b^2 = \dfrac{4\delta^2}{\Delta^3}$, which simplifies to $a^2b^2 = 4(s+t-1)^2 s^4 t^4 g(u)$, where

$$g(u) = \frac{u^2(su+t)^2}{I^3(u)}.$$

After some simplification, $g'(u) = \dfrac{2u(su+t)q(u)}{I^4(u)}$, where

$$q(u) = s^3 (s-1)^2 u^3 + s^2 t(2(s-1)^2 + st + s + t - 1)u^2$$
$$-st^2(2(t-1)^2 + st + s + t - 1)u - t^3 (t-1)^2.$$

Since $0 \leq u_1 < u, u(su+t) \neq 0$, and thus $g'(u) = 0 \iff q(u) = 0$. Recalling that $(s,t) \in G$ (see (1.2)), the coefficients of u^3 and of u^2 in $q(u)$ are positive, while the coefficients of u and the constant term in $q(u)$ are negative. Hence q has precisely one sign change, which implies that q has exactly one real root in $(0, \infty)$ by Descartes' Rule of Signs. Thus g has precisely one critical point in (u_1, u_2). Since u_1 and u_2 are the roots of $I, \lim_{u \to u_1^+} g(u) = \lim_{u \to u_2^-} g(u) = \infty$, which implies that g must attain its global minimum on (u_1, u_2). ∎

Remark 8.4 *In [38] and [39], the authors investigate the problem of constructing and characterizing an ellipse of minimal area containing a finite set of points. The results and methods used here are different than those in [38] and [39], but it is worth pointing out some of the small intersection. In particular, for a convex quadrilateral, Q, the authors in [38] and [39] construct an algorithm for finding the minimal area ellipse containing Q and they also prove a uniqueness result. For the case when this ellipse passes thru all four vertices of Q, this ellipse is then the minimal area ellipse discussed in this paper. However, there are convex quadrilaterals, Q, for which the minimal area ellipse containing Q does not pass thru all four vertices of Q. In that case, the the minimal area ellipse discussed in this paper is not the same.*

8.4   Examples

(1) Consider the quadrilateral $Q_{4,2}$. Using (8.1) with $s = 4$, $t = 2$, $v = 1$, and $w = 0$, the general equation of any ellipse circumscribed about $Q_{4,2}$ is given by

$$4ux^2 - (6u+1)xy + 4y^2 - 4ux - 4y = 0,$$

$$\frac{13 - 4\sqrt{10}}{18} < u < \frac{13 + 4\sqrt{10}}{18}. \tag{8.19}$$

Using (8.2), the center of any ellipse circumscribed about $Q_{4,2}$ is

$$\left(\frac{-4(1+14u)}{36u^2 - 52u + 1}, \frac{-12u(3+2u)}{36u^2 - 52u + 1} \right). \tag{8.20}$$

(a) We now find the ellipse of minimal area, E_A, circumscribed about $Q_{4,2}$. By the proof of Theorem 8.3, the equation of E_A is given by (8.19) with $u = u_A$, where u_A is the unique root of $q(u) = 8(72u^3 + 124u^2 - 30u - 1)$ in $\left(\frac{13 - 4\sqrt{10}}{18}, \frac{13 + 4\sqrt{10}}{18} \right) \approx (0.01\,95, 1.\,425\,0)$. Now $u_A \approx 0.241\,47$, so letting $u = 0.241\,47$ in (8.19) yields the approximate equation of E_A: $0.965\,88x^2 - 2.\,448\,8xy + 4y^2 - 0.965\,88\, x - 4y = 0$. To find the approximate center of E_A, let $u = 0.241\,47$ in (8.20), which gives $(1.\,852\,8, 1.\,067\,1)$. It is easy to show that the center of E_A lies in int $(Q_{4,2})$. However, that is not always the case, as we show now.

(b) Letting $u = 1$ in (8.20) yields $(4, 4) \notin$ int $(Q_{4,2})$. Thus the center of an ellipse circumscribed about Q does not always lie inside Q! We will see below that that is not the case for parallelograms, however.

(2) Let Q be the quadrilateral with vertices $(-3, -1), (-1, 4), (2, 6)$, and $(4, 1)$.

(a) Find the general equation of any ellipse circumscribed about Q.

(b) Find the ellipse of minimal eccentricity, E_O, circumscribed about Q.

We considered this quadrilateral earlier when doing inscribed ellipses, where we noted that $\mathbb{F}_{s,t,v,w}(Q) = Q_{s,t,v,w}$, where $\mathbb{F}_{s,t,v,w}(x, y) = \frac{1}{29}(5x - 2y + 13, 2x + 5y + 11)$ and where $s = \frac{11}{29}$, $t = \frac{45}{29}$, $v = \frac{31}{29}$, and $w = \frac{24}{29}$. Then $f_1 = \frac{19}{29}$, $f_2 = \frac{39}{29}$, $f_3 = \frac{7161}{24\,389}$, and $f_4 = -\frac{6820}{24\,389}$.

(a) Substituting into (8.1) and simplifying gives

$$7161ux^2 + 20(341u - 1182)xy + 7161y^2 +$$
$$3(6840 - 4433u)x - 7161y = 0, \tag{8.21}$$
$$1.\,182\,6 < u < 10.\,160,$$

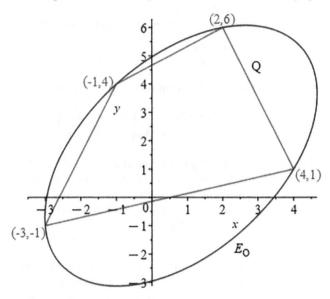

FIGURE 8.1
Ellipse of minimal eccentricity circumscribed about Q

where the bounds on u are decimal approximations using the formulas for u_1 and u_2 from Lemma 8.1(i).

(b) Using (8.4), the minimal eccentricity occurs for $u = u_0 = \dfrac{489\,081}{184\,481}$. Substituting that value for u into (8.21) yields the equation below for the ellipse, $E_{s,t,v,w}$, of minimal eccentricity circumscribed about $Q_{s,t,v,w}$ with the values of s,t,v,w given above: $\Psi(x,y) = 0$, where

$$\Psi(x,y) = 489\,081x^2 - 143\,220xy + \qquad (8.22)$$
$$184\,481y^2 - 379\,659x - 184\,481y.$$

The equation of E_O is then given by $\Psi(\mathsf{F}_{s,t,v,w}(x,y)) = 0$, which yields the equation

$$397\,681x^2 - 313\,780xy + 275\,881y^2 - 194\,951x - 552\,665y = 4051\,188.$$

9

Parallelograms

9.1 Equation

Recall the parallelogram, $Q_{d,k,l}$, with vertices $(-l-d, -k), (-l+d, k), A(l+d, k)$, and $(l-d, -k)$, where $l, k > 0, 0 \le d < l$. Also recall that $Q_{d,k,l}$ is a rectangle if and only if $d = 0$.

Proposition 9.1 E_0 *is an ellipse passing thru the vertices of* $Q_{d,k,l}$ *if and only if the general equation of* E_0 *is given by*

$$k(u - k^2)x^2 - 2d(u - k^2)xy + k\left(l^2 - d^2\right)y^2 - k\left(l^2 - d^2\right)u = 0,$$

$$k^2 < u < \frac{k^2 l^2}{d^2} \text{ if } d \ne 0, \ k^2 < u < \infty \text{ if } d = 0. \tag{9.1}$$

In addition, the center of E_0 *is* $(0,0)$.

Proof. Suppose that E_0 is an ellipse passing thru the vertices of $Q_{d,k,l}$ and let $P(x,y) = a_6 x^2 + a_5 xy + a_4 y^2 + a_3 x + a_2 y + a_1$; Assume first that $d \ne 0$. If P vanishes at the vertices of $Q_{d,k,l}$, we have $P(A_1) + P(A_3) - P(A_2) - P(A_4) = 0 \Rightarrow 8a_6 dl + 4ka_5 l = 0 \Rightarrow 2a_6 d + ka_5 = 0 \Rightarrow a_6 = -\dfrac{k}{2d}a_5$; $P(A_1) + P(A_4) - P(A_2) - P(A_3) = 0 \Rightarrow -4a_3 d - 4ka_2 = 0 \Rightarrow a_3 d + ka_2 = 0$; $P(A_1) + P(A_2) - P(A_4) - P(A_3) = 0 \Rightarrow -4a_3 l = 0 \Rightarrow a_3 = 0$, which implies that $a_2 = 0$ as well. Now $P(A_1) = (d+l)^2 a_6 + k(l+d)a_5 + k^2 a_4 - (l+d)a_3 - ka_2 + a_1 = 0 \Rightarrow (d+l)^2\left(-\dfrac{k}{2d}a_5\right) + k(l+d)a_5 + k^2 a_4 + a_1 = 0 \Rightarrow k\dfrac{d^2 - l^2}{2d}a_5 + k^2 a_4 + a_1 = 0 \Rightarrow a_5 = 2d\dfrac{k^2 a_4 + a_1}{k\left(l^2 - d^2\right)}$, which implies that $a_6 = -\dfrac{k}{2d}\left(2d\dfrac{k^2 a_4 + a_1}{k\left(l^2 - d^2\right)}\right) = \dfrac{k^2 a_4 + a_1}{d^2 - l^2}$; Hence $k\left(d^2 - l^2\right)P(x,y) = k(k^2 a_4 + a_1)x^2 - 2(k^2 a_4 + a_1)dxy + a_4 k\left(d^2 - l^2\right)y^2 + a_1 k\left(d^2 - l^2\right)$. If $a_4 = 0$, it is not hard to show that $G(x,y) = 0$ does not yield an ellipse. Thus we may assume that $a_4 \ne 0$. Now divide thru by $-a_4$ and let $u = -\dfrac{a_1}{a_4}$ to get the left hand side of the equation in (9.1).

Thus $P(x,y) = Ax^2 + Bxy + Cy^2 + Dx + Ey + F$, where

$$A = k(u - k^2), B = -2d(u - k^2), C = k\left(l^2 - d^2\right),$$
$$D = 0, E = 0, F = -k\left(l^2 - d^2\right)u. \tag{9.2}$$

DOI: 10.1201/9781003474890-11

$d < l$ and $k > 0 \Rightarrow C > 0$. Since we have assumed that E_0 is an ellipse, that implies that $A > 0$. Also, by Lemma A.1 in the Appendix, $\Delta = 4AC - B^2 = 4\left(u - k^2\right)\left(k^2 l^2 - ud^2\right) > 0$. Since $A > 0$, $k^2 l^2 - ud^2 > 0$, which gives the bounds on u in (9.1). Now suppose that E_0 has equation $P(x, y) = 0$ with $k^2 < u < \dfrac{k^2 l^2}{d^2}$ and where the coefficients of $P(x, y)$ are given by (9.2). It is simple to prove that $P(A_j) = 0, j = 1\text{–}4$, which implies that E_0 passes thru the vertices of $Q_{d,k,l}$. As above, it follows that $A, C > 0$ and $\Delta = 4AC - B^2 > 0$. Also, $\delta = CD^2 + AE^2 - BDE - F\Delta = 4k\left(l^2 - d^2\right) u\left(u - k^2\right)\left(k^2 l^2 - ud^2\right) > 0$, which implies that $P(x, y) = 0$ defines the equation of an ellipse. The case $d = 0$ follows easily as above and we omit the details. That proves (9.1). Using (9.2) and Lemma A.2, it follows easily that the center of E_0 is $(0, 0)$. ∎

Remark 9.1 *As in the proof of Proposition 2.1 for ellipses inscribed in $Q_{d,k,l}$, we could first have proven Proposition 9.1 for the square, S. However, that doesn't really simplify things much.*

Using Proposition 9.1 we prove the following probably known fact.

Corollary 5 *A parallelogram is cyclic if and only if it is a rectangle.*

Proof. We may assume that the parallelogram equals $Q_{d,k,l}$ for some $l, k > 0$; $0 \le d < l$. If $Q_{d,k,l}$ is cyclic, then (9.1) defines the equation of a circle. Now if $d \ne 0$, then $u - k^2 > 0$, which implies that the coefficient of xy in (9.1) does not equal 0. But then (9.1) does not define the equation of a circle, and hence $d = 0$, which implies that $Q_{d,k,l}$ is a rectangle. Conversely, if $Q_{d,k,l}$ is a rectangle, then $d = 0$ and the circle $x^2 + y^2 = k^2 + l^2$ passes thru the vertices of $Q_{d,k,l}$. ∎

9.2   Minimal Eccentricity

Theorem 9.1 *Let Q be a parallelogram. Then there is a unique ellipse of minimal eccentricity passing thru the vertices of Q.*

Proof. As usual, we may assume that Q is **not** a cyclic quadrilateral. By Lemma 2.4 we may assume that $Q = Q_{d,k,l}$ for some $l, k > 0$, $0 \le d < l$. Let E_0 be an ellipse passing thru the vertices of $Q_{d,k,l}$ and let a and b denote the lengths of the semi–major and semi–minor axes, respectively, of E_0. Let A–C be given by (9.2). Using Lemma A.1 in the Appendix and simplifying yields

$$\frac{b^2}{a^2} = Z(u) = \frac{k\left(u + l^2 - d^2 - k^2\right) - \sqrt{H(u)}}{k\left(u + l^2 - d^2 - k^2\right) + \sqrt{H(u)}}, \tag{9.3}$$

$$H(u) = k^2\left((u - k^2) - (l^2 - d^2)\right)^2 + 4d^2(u - k^2)^2.$$

Note that $k^2 < u \Rightarrow u + l^2 - d^2 - k^2 > 0$ since $l > d$. Thus $k\left(u + l^2 - d^2 - k^2\right) > 0$; Also, $H(u) \neq 0$ since we assumed that Q is not cyclic. Thus $Z(u)$ is differentiable on $k^2 < u < \dfrac{k^2 l^2}{d^2}$ if $d \neq 0$, $k^2 < u < \infty$ if $d = 0$. We want to maximize $Z(u)$. Proceeding as in the proof of Theorem 8.1, the numerator of $Z'(u)$ equals $0 \iff 2H(u) - \left(u + l^2 - d^2 - k^2\right)H'(u) = 0 \iff u = u_0$, where $u_0 = k^2\dfrac{l^2 + k^2 + d^2}{k^2 + 2d^2}$. Now $u_0 - k^2 = \dfrac{k^2\left(l^2 - d^2\right)}{k^2 + 2d^2} > 0$ and, if $d \neq 0, \dfrac{k^2 l^2}{d^2} - u_0 = \dfrac{k^2\left(k^2 + d^2\right)\left(l^2 - d^2\right)}{d^2\left(k^2 + 2d^2\right)} > 0$, which implies that $k^2 < u_0 < \dfrac{k^2 l^2}{d^2}$. If $d = 0$, then $u_0 = l^2 + k^2 \in (k^2, \infty)$. It follows easily that $Z(u_0)$ is the global maximum of $Z(u)$ on $\left(k^2, \dfrac{k^2 l^2}{d^2}\right), d \neq 0$, or on $\left(k^2, \infty\right)$ if $d = 0$. By Proposition 9.1, letting $u = u_0$ in (9.1) yields the equation of an ellipse. ∎

What follows now is in essence Proposition 8.2 for parallelograms.

Proposition 9.2 *If Q is a parallelogram, then there is only one pair of conjugate directions, m_1 and m_2, belonging to all of the ellipses which pass thru the vertices of Q. In addition, each ellipse which passes thru the vertices of Q possesses those conjugate direction pairs.*

Proof. By Lemma 2.4 we may assume that $Q = Q_{d,k,l}$ for some $l, k > 0$; $0 \leq d < l$. Suppose that E_0 is an ellipse which passes thru the vertices of $Q_{d,k,l}$. Since the center of any ellipse of circumscription is $(0,0)$ by Proposition 9.1, any conjugate diameter of E_0 must pass thru $(0,0)$. Let DI_1 be a diameter of E_0 with equation $y = mx_1$ and let DI_2 be a diameter of E_0 with equation $y = m'x$; Suppose that DI_1 intersects E_0 at the two distinct points $P_j = (x_j, y_j), j = 1, 2$ and let T_1 be the tangent line to E_0 at P_1. Differentiating both sides of (9.1) with respect to x and substituting $x = x_1, y = y_1 = mx_1$ yields slope of $T_1 = -\dfrac{2Ax_1 + Bmx_1 + D}{Bx_1 + 2Cmx_1 + E}$. Substituting for $A - E$ using (9.2) and proceeding as in the proof of Proposition 8.2, one can show that DI_1 and DI_2 are *conjugate* diameters if and only

$$m' = \frac{\left(u - k^2\right)\left(dm - k\right)}{mk(l^2 - d^2) - d\left(u - k^2\right)}. \tag{9.4}$$

Again we ignore for now the fact that certain values of u do not yield an ellipse. Assume first that $d \neq 0$. We want to find values of m and m' so that (9.4) holds for **all** values of $u, k^2 < u < \dfrac{k^2 l^2}{d^2}$; $u = k^2 \Rightarrow m' = 0$ and $u = \dfrac{k^2 l^2}{d^2} \Rightarrow m' = \dfrac{k}{d}$; Now substitute $m = \dfrac{k}{d}$ into the right hand side of (9.4). That trivially gives $m' = 0$ for any value of u, which implies that (9.4) holds with $m = 0$ and $m' = \dfrac{k}{d}$ for all values of u, and in particular for $k^2 < u < \dfrac{k^2 l^2}{d^2}$. Thus the only pair of conjugate directions belonging to all of the ellipses which pass thru the vertices of $Q_{d,k,l}$ are $m_1 = 0$ and $m_2 = \dfrac{k}{d}$. In addition, each ellipse which passes thru the vertices of $Q_{d,k,l}$ possesses the conjugate direction pairs m_1 and m_2. If $d = 0$, then E_0 has equation $(u-k^2)x^2+l^2y^2-l^2u = 0$ and it follows easily that the unique pair of conjugate directions, m_1 and m_2, belonging to all of the ellipses of circumscription are 0 and ∞. The corresponding conjugate diameters in this case are part of the x and y axes. ∎

What follows next is in essence Theorem 8.2 for parallelograms.

Theorem 9.2 *Let Q be a parallelogram which is **not** a rectangle. Let m_1 and m_2 be the unique pair of conjugate directions belonging to all of the ellipses of circumscription and let E_0 be any ellipse of circumscription. Then the equal conjugate diameters of E_0 possess the directional constants m_1 and m_2 if and only if E_0 is the unique ellipse of circumscription of minimal eccentricity.*

Proof. Again, by Lemma 2.4 we may assume that $Q = Q_{d,k,l}$ for some $l, k > 0$; $0 < d < l$. Let $m_1 = 0$ and $m_2 = \dfrac{k}{d}$ be the unique pair of conjugate directions belonging to all of the ellipses of circumscription given in Proposition 9.2. Suppose that E_0 is an ellipse which passes thru the vertices of $Q_{d,k,l}$, and let CD_1 and CD_2 be the conjugate diameters of E_0 which possess the directional constants m_1 and m_2, respectively. Note that any conjugate diameter must pass thru $(0,0)$. Then CD_1 has equation $y = 0$, and CD_2 has equation $y = \dfrac{k}{d}x$; Suppose that CD_1 intersects E_0 at the two distinct points $P_j = (x_j, 0), j = 1, 2$ and CD_2 intersects E_0 at the two distinct points $P_j = \left(x_j, \dfrac{k}{d}x_j\right), j = 3, 4$; CD_1 and CD_2 are **equal** conjugate diameters if and only if

$$(x_2 - x_1)^2 = (x_4 - x_3)^2 \left(1 + \frac{k^2}{d^2}\right). \tag{9.5}$$

Since $\dfrac{x_1 + x_2}{2} = \dfrac{x_3 + x_4}{2} = x$ coordinate of the center of E_0, $x_2 = -x_1$ and $x_4 = -x_3$ and thus (9.5) becomes $4x_1^2 = 4x_3^2\left(1 + \dfrac{k^2}{d^2}\right)$. Substituting x_1 for x and 0 for y in (9.1) implies that $x_1^2 = \dfrac{u\left(l^2 - d^2\right)}{u - k^2}$. Substituting x_3 for x and $\dfrac{k}{d}$ for y in (9.1) implies that $x_3^2 = \dfrac{ud^2(l^2 - d^2)}{k^2l^2 - d^2u}$. Hence $4x_1^2 = 4x_3^2\left(1 + \dfrac{k^2}{d^2}\right) \iff \dfrac{u\left(l^2 - d^2\right)}{(u - k^2)} = \dfrac{u\left(l^2 - d^2\right)\left(d^2 + k^2\right)}{k^2l^2 - ud^2} \iff \dfrac{1}{u - k^2} = \dfrac{d^2 + k^2}{k^2l^2 - ud^2} \iff u = u_0 = k^2\dfrac{l^2 + k^2 + d^2}{k^2 + 2d^2}$, the value of u which gives the unique ellipse of minimal eccentricity passing thru the vertices of $Q_{d,k,l}$ given in the proof of Theorem 9.1. ∎

Remark 9.2 *As noted in the remark after Theorem 8.2, Theorem 9.2 would not hold if Q is a cyclic quadrilateral. But a parallelogram is a cyclic quadrilateral if and only if it is a rectangle.*

9.3 Minimal Area

Theorem 9.3 *Let Q be a parallelogram. Then there is a unique ellipse of minimal area, $E_{\min}$, passing thru the vertices of Q. In addition, $\dfrac{Area\,(E_{\min})}{Area(Q)} = \dfrac{\pi}{2}$.*

Proof. Assume that $Q = Q_{d,k,l}$ and let E_0 denote an ellipse passing thru the vertices of $Q_{d,k,l}$ and let a and b denote the lengths of the semi–major and semi–minor axes, respectively, of E_0. Using (9.2), $\Delta = 4AC - B^2 = 4\left(u - k^2\right)\left(k^2 l^2 - ud^2\right)$ and $\delta = CD^2 + AE^2 - BDE - F\Delta = 4k\left(l^2 - d^2\right)u\left(u - k^2\right)\left(k^2 l^2 - ud^2\right)$. By Lemma A.2 in the Appendix, $a^2 b^2 = \dfrac{4\delta^2}{\Delta^3}$, which simplifies to $f(u) = \dfrac{k^2\left(l^2 - d^2\right)^2 u^2}{\left(u - k^2\right)\left(k^2 l^2 - ud^2\right)}$; $f'(u) = k^4\left(l^2 - d^2\right)^2 u\dfrac{(l^2 + d^2)u - 2k^2 l^2}{\left(u - k^2\right)^2\left(-k^2 l^2 + ud^2\right)^2} = 0 \iff u = \dfrac{2k^2 l^2}{l^2 + d^2}$; $\lim_{u \to k^2+} f(u) = \infty$ and $\lim_{u \to k^2 l^2 / d^2-} f(u) = \infty$ if $d \neq 0$; If $d = 0$, then $f(u) = \dfrac{u^2}{u - k^2}$, which implies that $\lim_{u \to \infty} f(u) = \infty$; Given the bounds on u in (9.1), $f\left(\dfrac{2k^2 l^2}{l^2 + d^2}\right) = 4k^2 l^2$ must be the unique global minimum of f, which proves existence/uniqueness. To finish the proof, we may assume that $Q = S$, the square with vertices $(-1, -1), (-1, 1), (1, 1)$, and $(1, -1)$. Using $d = 0$ and $k = l = 1$, the global minimum of f equals 4, which implies that Area$(E_{\min}) = 2\pi$. Thus $\dfrac{\text{Area}\,(E_{\min})}{\text{Area}(Q)} = \dfrac{2\pi}{4} = \dfrac{\pi}{2}$. ∎

Remark 9.3 *We could have assumed that Q = S for the first part of the proof of Theorem 9.3. However, we use the value of u where the minimal area is attained for $Q_{d,k,l}$ in general in the proof of Theorem 9.7 below.*

9.4 Example

By Proposition 9.1, with $d = 1$, $k = 4$, and $l = 3$, the equation of any ellipse passing thru the vertices of $Q_{1,4,3}$ is $2\left(u - 16\right)x^2 - \left(u - 16\right)xy +$

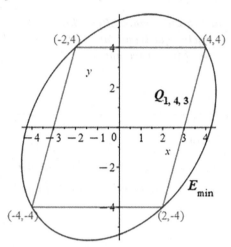

FIGURE 9.1
Circumscribed ellipse of minimal area

$16y^2 = 16u, 16 < u < 144$. By Theorem 9.3 and its proof, the unique ellipse, E_{min}, of minimal area passing thru the vertices of $Q_{1,4,3}$ is attained when $u = \frac{144}{5}$ and thus E_{min} has equation $8x^2 - 4xy + 5y^2 = 144$.

By Theorem 9.1 and its proof, the unique ellipse, E_O, of minimal eccentricity passing thru the vertices of $Q_{1,4,3}$ is attained when $u = \frac{208}{9}$ and thus E_O has equation $8x^2 - 4xy + 9y^2 = 208$.

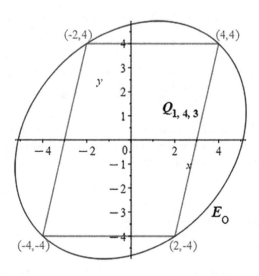

FIGURE 9.2
Circumscribed ellipse of minimal eccentricity

We saw in Theorem 2.4 in § 2.4 that the ellipse of maximal area, the ellipse of minimal eccentricity, and the ellipse of maximal arc length **inscribed** in a **rectangle** are all the same. What about ellipses **circumscribed** about rectangles? Consider the rectangle, Z, with vertices $(-l, -k), (-l, k), (l, k)$, and $(l, -k)$, where $l, k > 0$. By (9.2) with $d = 0$, the equation of any ellipse circumscribed about Z is given by $(u - k^2)x^2 + l^2y^2 - l^2u = 0, k^2 < u < \infty$. By the proof of Theorem 9.3, it follow easily that $u = 2k^2$ gives the ellipse of minimal area, E_{min}, circumscribed about Z. The equation of E_{min} is then $k^2x^2 + l^2y^2 = 2k^2l^2$. By the proof of Theorem 9.1, $u = l^2 + k^2$ gives the ellipse of minimal eccentricity, E_O, circumscribed about Z. The equation of E_O is then $x^2 + y^2 = l^2 + k^2$. Thus $E_{min} \neq E_O$ if $k \neq l$ since E_O is a circle and E_{min} is not a circle. In other words, for non-square rectangles, the ellipse of circumscription of minimal area and the ellipse of circumscription of minimal eccentricity are not the same. It is clear from above that they are the same for squares. What about arc length? Even for squares, this seems to be a fairly non-trivial problem (at least so far for this author). The numerical evidence seems to indicate the following:

Conjecture 9.4 *Let Z be a rectangle. Then there is a unique ellipse of minimal arc length circumscribed about Z and that ellipse equals the ellipse of minimal eccentricity circumscribed about Z.*

Note that there does not appear to be an ellipse of **maximal** arc length circumscribed about a rectangle.

9.5   Area Inequality

In this section we prove an area inequality for ellipses passing thru the vertices of a quadrilateral which is similar to Theorem 3.1 for ellipses inscribed in quadrilaterals.

In Theorem 3.1 in § 3 we proved an inequality for circles inscribed in tangential quadrilaterals. The following inequality is the analog of Theorem 3.1 for circles circumscribed about cyclic quadrilaterals.

Proposition 9.3 *Let Q be a cyclic quadrilateral, let $A(Q)$ denote its area, let C_0 be it's circumscribed circle, and let r denote the circumradius (radius of C_0). Then $A(Q) \leq 2r^2$, with equality if and only if Q is a square.*

Proof. The proof that $A(Q) \leq 2r^2$ for any cyclic quadrilateral, Q, is given in [40], Lemma 4.4. The proof given also shows that equality holds if and only if Q is a square. That also follows from [31], Theorem 7.1(b). Proposition 9.3 is also essentially given in the statement of Theorem 7.1(c) in [31], but no proof is given. ■

Theorem 9.5: *Let E_0 be any ellipse passing thru the vertices of a convex quadrilateral, Q, and let $A(E_0)$ denote its area. Then $A(E_0) \geq \frac{\pi}{2}A(Q)$, and equality holds if and only if Q is a parallelogram.*

Proof. Use an affine transformation, $\mathbb{F}$ (for example, an orthogonal projection as done in [20]) to map E_0 to a circle, C_0, passing thru the vertices of the cyclic quadrilateral $Q_1 = \mathbb{F}(Q)$; Since parallel lines are preserved under affine transformations, Q_1 is a square if and only if Q is a parallelogram, Also, as done earlier in this book, since ratios of areas of ellipses are preserved under affine transformations, $\dfrac{A(E_0)}{A(Q)} = \dfrac{A(C_0)}{A(Q_1)} = \dfrac{\pi R^2}{A(Q_1)}$; Suppose first that Q is **not** a parallelogram. Then Q_1 is **not** a square and by Proposition 9.3, $\dfrac{\pi r^2}{A(Q_1)} > \dfrac{\pi r^2}{2r^2} = \dfrac{\pi}{2}$, which implies that $\dfrac{A(E_0)}{A(Q)} > \dfrac{\pi}{2}$. Now suppose that Q is a parallelogram. Then equality holds by Theorem 9.3. ∎

Part III

Inscribed versus Circumscribed

In this part we relate and compare some of the material from Parts 1 and 2. First we prove a result comparing the centers of ellipse inscribed in and circumscribed about a given convex quadrilateral (Theorem 9.6). We then prove a result comparing the ellipses of maximum area inscribed in and circumscribed about a given parallelogram (Theorem 9.7). Then we discuss a category of quadrilaterals we call bielliptic, which is a generalization of bicentric quadrilaterals–see § 10.

We have seen that ellipses inscribed in and ellipses circumscribed about parallelograms always have the same center. The following theorem shows that parallelograms are the **only** quadrilaterals for which this holds.

Theorem 9.6 *Let Q be a convex quadrilateral in the xy plane which is **not** a parallelogram. Suppose that E_1 and E_2 are each ellipses, with E_1 inscribed in Q and E_2 circumscribed about Q. Then E_1 and E_2 cannot have the same center.*

Proof. Since the center of an ellipse is affine invariant, we may assume that $Q = Q_{s,t}$ for some $(s,t) \in G$ (see 1.2). By Proposition 8.1, with $v = 1$ and $w = 0$, the center of E_2 equals (x_0, y_0), where $x_0 = \dfrac{st(s(s+2t-1)u + t^2 - t)}{I(u)}$, $y_0 = \dfrac{stu(s(s-1)u + t(2s+t-1))}{I(u)}$, and $I(u)$ is given by (8.18). The center of E_1 must lie on the Newton line $y = L(x) = \dfrac{1}{2}\dfrac{s-t+2x(t-1)}{s-1}$ (see 1.4), which implies that $L(x_0) - y_0 = 0$; Now $L(x_0) - y_0$
$$= \frac{1}{2}\frac{(s+t)\left((t^2-t)^2 - (s^2-s)^2 u^2\right)}{(s-1)I(u)} = 0 \iff u^2 = \frac{(t^2-t)^2}{(s^2-s)^2} \iff u = \left|\frac{t^2-t}{s^2-s}\right|.$$
As noted in § 1.1, the x–coordinate of the center of any ellipse inscribed in $Q_{s,t}$ must lie in the open interval, I, with $\dfrac{1}{2}s$ and $\dfrac{1}{2}$ as endpoints(see 1.3). If $u = \dfrac{t^2-t}{s^2-s}$, then $x_0 = \dfrac{1}{2}s \notin I$; If $u = -\dfrac{t^2-t}{s^2-s}$, then $x_0 = \dfrac{1}{2} \notin I$; Thus E_1 and E_2 cannot have the same center. ■

DOI: 10.1201/9781003474890-12

We now prove a result comparing the ellipse of maximum area, $E_{\max}$, inscribed in a given parallelogram, Q, with the ellipse of minimum area, $E_{\min}$, passing thru the vertices of Q.

Theorem 9.7 *Let Q be a parallelogram. Then $E_{\max}$ and $E_{\min}$ are confocal ellipses.*

Proof. Assume that $Q = Q_{d,k,l}$ for some $l, k > 0$; $0 \le d < l$. For $E_{\max}$, using the coefficients in (2.4) from Proposition 2.1, $\Delta = 4k^2l^2(1 - u^2)$ and $\delta = 4k^4l^4(1 - u^2)^2$; By Lemma 12.2 in the Appendix, $a^2b^2 = \dfrac{4\delta^2}{\Delta^3}$, which simplifies to $k^2l^2\left(1 - u^2\right)$; Thus a^2b^2 attains its maximum when $u = 0$, which, upon substituting into (2.4) gives $k^2x^2 - 2dkxy + (d^2 + l^2)y^2 = k^2l^2$ for the equation of $E_{\max}$. By the proof of Theorem 9.3, letting $u = \dfrac{2k^2l^2}{l^2 + d^2}$ in (9.1) yields the equation of $E_{\min}$, which simplifies to $k^2x^2 - 2dkxy + (d^2 + l^2)y^2 = 2k^2l^2$. Hence $E_{\max}$ and $E_{\min}$ are confocal ellipses. ∎

Remark 9.4 *Theorem 9.7 does **not** hold in general if $E_{\max}$ and $E_{\min}$ are replaced by the ellipses of minimal **eccentricity** inscribed in, or passing thru the vertices of, the parallelogram, Q.*

Remark 9.5 *Theorem 9.7 implies that Theorem 2.2(iii), where $E_A = E_{\max}$, also holds for ellipses passing thru the vertices of a parallelogram, Q. That is, if P is the polynomial with roots at the vertices of Q, then the three roots of $P'(z)$ are the foci and center of $E_{\min}$. Theorem 2.6 also holds as well for $E_{\min}$.*

10

Bielliptic Quadrilaterals

The following definition is well-known: A convex quadrilateral is called **bicentric** if it is both cyclic and tangential. We generalize the notion of bicentric quadrilaterals as follows. Let E_I be the unique ellipse of minimal eccentricity inscribed in a given convex quadrilateral, Q, and let E_O be the unique ellipse of minimal eccentricity circumscribed about Q.

Definition 8 *A convex quadrilateral is called **bielliptic** if E_I and E_O have the same eccentricity.*

If Q is bielliptic, we say that Q is of class α, $0 \leq \alpha < 1$, if E_I and E_O each have eccentricity α. Below we show that there do exist bielliptic parallelograms of class α for **each** $0 < \alpha < 1$ (see Theorem 10.3). We are also able to characterize bielliptic parallelograms by showing that a parallelogram is bielliptic if and only if the square of the length of one of the diagonals equals twice the square of the length of one of the sides (see Theorem 10.2). The latter result was proven in [14]. Note that a parallelogram is bicentric if and only if it is a square (see [31], Theorem 4.1(g)). This is not the case for bielliptic parallelograms as Theorem 10.2 shows.

For quadrilaterals in general, we are able to prove that there exists a convex quadrilateral, Q, which is not a parallelogram and which is bielliptic of class α for **some** $\alpha > 0$ (see Theorem 10.1). This result was proven in [16], but there is a small gap there in the proof. The proof here is a little different than in [16] and we fill in some details which were left out there. Our first main result is:

Theorem 10.1 *There exists a convex quadrilateral, Q, which is not a parallelogram and which is bielliptic of class α for some $\alpha > 0$. That is, there exists a bielliptic convex quadrilateral which is not a parallelogram and which is not bicentric.*

The notation we use here emphasizes the dependence on s and on t: Let $E_I(s,t)$ and $E_O(s,t)$ denote the inscribed and circumscribed ellipses of minimal eccentricity, respectively, for $Q_{s,t}$, $(s,t) \in G$. Let $\min_I(s,t) = $ eccentricity of $E_I(s,t)$ and let $\min_O(s,t) = $ eccentricity of $E_O(s,t)$. We shall specify a value of s, say s_0, and then show that $Q_{s_0,t}$ is bielliptic, but not bicentric, for some t (one could also specify a value of t as well). Before proving Theorem 10.1, we prove the following lemmas. The proofs given are for $s_0 = \dfrac{1}{2}$, though the proofs work for any specified value of s.

DOI: 10.1201/9781003474890-13

Lemma 10.1 $\min_I \left(\frac{1}{2}, t \right)$ *is a continuous function of t for $t > 0$.*

Proof. Let $a(t)$ and $b(t)$ denote the lengths of the semi–major and semi–minor axes, respectively, of $E_I \left(\frac{1}{2}, t \right)$. It suffices to prove that $\frac{b^2}{a^2}(t)$ is a continuous function of t. Define the following functions of $r \in J = (0, 1)$ as done in the proof of Theorem 1.4, but here with $v = 1, w = 0$, and $s = \frac{1}{2}$. We use the subscript t to emphasize that these functions also vary with t: $A_t(r) = \frac{1}{4}(2tr - r - 2t)^2$, $B_t(r) = -r^2 + \left(t + \frac{3}{2} \right)r - t$, $C_t(r) = \frac{1}{4}$, $O_t(r) = A_t(r) + C_t(r)$, $M_t(r) = (A_t(r) - C_t(r))^2 + (B_t(r))^2$, $p_t(r) = 2M_t(r)O'_t(r) - O_t(r)M'_t(r)$, and $N_t(r) = O_t^2(r) - M_t(r)$. From the proof of Theorem 1.4 , we have $\frac{b^2}{a^2}(t) = G_t(r) = \frac{N_t(r)}{\left(O_t(r) + \sqrt{M_t(r)} \right)^2}$. Also from the proof of Theorem 1.4, $M_t(r) > 0$ on J, which implies that $G_t(r)$ is differentiable on J and $G'_t(r) = \frac{p_t(r)}{\left(O_t(r) + \sqrt{M_t(r)} \right)^2 \sqrt{M_t(r)}}$. In addition, the 4th degree polynomial p_t has a unique root in J, which is the unique critical point of $G_t(r)$ in J and which we denote by $r_I(t)$. Thus $\min_I \left(\frac{1}{2}, t \right) = 1 - \sqrt{G_t(r_I(t))}$. Now $p_t(r) = a_4(t)r^4 + a_3(t)r^3 + a_2(t)r^2 + a_1(t)r + a_0(t)$, where $a_4(t) = -\frac{1}{4}(2t + 1)(2t - 1)(2t - 3)$–it is not necessary to give the other coefficient polynomials $a_j(t), j = 0$–3. Note that Theorem 1.4 assumes that (i) Q is **not** a trapezoid, and the **proof** of Theorem 1.4 given with the functions of r above assumes that (ii) Q is **not** a tangential quadrilateral. Now $Q_{s,t}$ is a trapezoid if and only if $t = 1$ and $Q_{s,t}$ is a tangential quadrilateral if and only if $s = t$. We shall consider those cases as limiting cases. Let $H = \left\{ t > 0 : t \neq 1, t \neq \frac{1}{2} \right\}$. Now $r_I(t)$ is clearly a continuous function of t on H, which implies that $\min_I \left(\frac{1}{2}, t \right)$ is a continuous function of t on H.

- If $t = \frac{1}{2}$, then $\lim_{t \to 1/2} p_t(r) = p_{1/2}(r) = (2r^2 - 4r + 1)(1 - r)$, and the unique root of $p_{1/2}(r)$ in J is $r_I \left(\frac{1}{2} \right) = 1 - \frac{1}{\sqrt{2}}$. To shorten notation here, use $r_1 = 1 - \frac{1}{\sqrt{2}}$ and $r_2 = 1 + \frac{1}{\sqrt{2}}$, the other root of $2r^2 - 4r + 1$; Now $\lim_{t \to 1/2} O_t(r) = O_{1/2}(r) = \frac{1}{2}, \lim_{t \to 1/2} M_t(r) = M_{1/2}(r) = \frac{1}{4}(2r^2 - 4r + 1)^2 = ((r - r_1)(r - r_2))^2$, and $\lim_{t \to 1/2} N_t(r) = N_{1/2}(r)$

$= -r\,(r-2)\,(r-1)^2$; Since the roots of a polynomial are continuous functions of the coefficients, $\lim\limits_{t \to 1/2} r_I(t) = r_I\left(\dfrac{1}{2}\right) = r_1$. Note that the latter limit holds even though the leading coefficient of $p_t(r)$, $a_4(t)$, approaches 0 as $t \to \dfrac{1}{2}$. Since $M_{1/2}(r_1) = 0$, $G_{1/2}$ is not differentiable on J. But $\lim\limits_{t \to 1/2} G_t(r) = G_{1/2}(r) = \dfrac{-r\,(r-2)\,(r-1)^2}{(\dfrac{1}{2}+|(r-r_1)(r-r_2)|\,)^2}$ and it follows(we do not provide all of the details here) that $\lim\limits_{t \to 1/2} G_t(r_I(t)) =$

$G_{1/2}(r_I\left(\dfrac{1}{2}\right)) = G(r_1) = -4r_1\,(r_1-2)\,(r_1-1)^2 = 1$. Since $Q_{1/2,1/2}$ is a tangential quadrilateral, $\dfrac{b^2}{a^2}(t) = 1$ when $t = \dfrac{1}{2}$ and thus $\min_I\left(\dfrac{1}{2},t\right)$ is continuous at $t = \dfrac{1}{2}$.

- If $t = 1$, then $\lim\limits_{t \to 1} p_t(r) = p_1(r) = \dfrac{1}{4}\,(r-2)\,q(r)$, where $q(r) = 3r^3 - 18r^2 + 27r - 10$; Now $q(0) < 0$, $q(1) > 0$, $q(2) < 0$, and $q(4) > 0$, which implies that q, and hence p, has a unique root $r_I(1) \in J$; $\lim\limits_{t \to 1} O_t(r) = O_1(r) = \dfrac{1}{4}(r^2 - 4r + 5)$, $\lim\limits_{t \to 1} M_t(r) = M_1(r) = \dfrac{1}{16}(17r^4 - 88r^3 + 154r^2 - 104r + 25)$, and $\lim\limits_{t \to 1} N_t(r) = N_1(r) = r\,(1-r)\,(r-2)^2$. One could show directly that $M_1(r)$ does not vanish on J, but instead it's easier to just follow the argument used in the proof of Lemma 1.7. If $M(r_0) = 0$ for some $r_0 \in J$, then $A(r_0) - C(r_0) = 0$ and $B(r_0) = 0$, which implies that the ellipse inscribed in $Q_{1/2,1}$ corresponding to r_0 is a circle. But that contradicts the fact that $Q_{1/2,1}$ is not a tangential quadrilateral. Thus $G_1(r)$ is differentiable on J. As noted above, since the roots of a polynomial are continuous functions of the coefficients, $\lim\limits_{t \to 1} r_I(t) = r_I\,(1)$. Thus $\lim\limits_{t \to 1} \dfrac{b^2}{a^2}(t) = \lim\limits_{t \to 1} G_t(r_I(t)) = G(r_I(1))$ and it follows that $\min_I\left(\dfrac{1}{2},t\right)$ is continuous at $t = 1$. ∎

Lemma 10.2 $\min_O\left(\dfrac{1}{2},t\right)$ *is a continuous function of t for $t > 0$.*

Proof. Let $a(t)$ and $b(t)$ denote the lengths of the semi–major and semi–minor axes, respectively, of $E_O\left(\dfrac{1}{2},t\right)$. As above, it suffices to prove that $\dfrac{b^2}{a^2}(t)$ is a continuous function of t. Define the following functions of u as done in the proof of Theorem 8.1, but here with $v = 1, w = 0$, and

$s = \dfrac{1}{2}$. Again we use the subscript t to emphasize that these functions also

vary with t: $H_t(u) = \dfrac{4t^2(u-1)^2 + (u - 4t\,(t-1)\,)^2}{16}$ and $K_t(u) = 2H_t(u) -$

$(u+1)H_t'(u)$. From the proof of Theorem 8.1 , we have $\dfrac{b^2}{a^2}(t) = Z_t(u) =$

$\dfrac{t^2(u+1)^2 - 4H_t(u)}{(t(u+1) + 2\sqrt{H_t(u)})^2}$. Let $u_O(t) = $ unique root of $K_t(u)$ in the open inter-

val $(u_1(t), u_2(t))$, where $u_1(t)$ and $u_2(t)$ are the negative and positive roots, repectively, of $4t\left(3t - 1 \pm \sqrt{2}\sqrt{t}\sqrt{4t - 2}\right)$. From (8.4) with $v = 1, w = 0$, and

$s = 1/2$ we have $u_O(t) = 4t\dfrac{4t^3 - 8t^2 + 7t - 1}{12t^2 - 4t + 1} = $ the unique critical point of

$Z_t(u)$ in $(u_1(t), u_2(t))$, which implies that $\min_O\left(\dfrac{1}{2}, t\right) = 1 - \sqrt{Z_t(u_O(t))}$.

Note that the proof of Theorem 8.1 given with the functions of u above assumes that Q is **not** a cyclic quadrilateral. Now it is easy to show that $Q_{s,t}$ is cyclic if and only

$$\left(s - \frac{1}{2}\right)^2 + \left(t - \frac{1}{2}\right)^2 = \frac{1}{2}. \tag{10.1}$$

If $s = \dfrac{1}{2}$, then since we want $t > 0$, (10.1) holds $\Longleftrightarrow t = t_1$, where

$$t_1 = \frac{\sqrt{2} + 1}{2}.$$

We shall consider the case $t = t_1$ as a limiting case. So let $L = \{t > 0 : t \neq t_1\}$. Now $u_O(t)$ is clearly a continuous function of t on H, which implies that $\min_O\left(\dfrac{1}{2}, t\right)$ is a continuous function of t on L. To prove continuity at t_1:

$\lim\limits_{t \to t_1} H_t(u) = H_{t_1}(u) = \dfrac{1}{8}\left(2 + \sqrt{2}\right)(u-1)^2$, which implies that

$$\lim_{t \to t_1} Z_t(u) = Z_{t_1}(u) =$$

$$\frac{t_1^2(u+1)^2 - 4\dfrac{1}{8}\left(2 + \sqrt{2}\right)(u-1)^2}{(t_1(u+1) + 2\sqrt{\dfrac{1}{8}\left(2 + \sqrt{2}\right)}|u - 1|)^2}.$$

Since $\lim\limits_{t \to t_1} u_O(t) = 1$ it follows(again we do not provide all of the details here) that $\lim\limits_{t \to t_1} Z_t(u_O(t)) = Z_{t_1}(u_O(t_1)) = Z_{t_1}(1) = 1$. Since $Q_{1/2, t_1}$ is a cyclic quadrilateral, $\dfrac{b^2}{a^2}(t) = 1$ when $t = t_1$ and thus $\min_O\left(\dfrac{1}{2}, t\right)$ is continuous at $t = t_1$. ∎

Proof. of Theorem 10.1: Let $f(t) = \min_O \left(\frac{1}{2}, t\right) - \min_I \left(\frac{1}{2}, t\right)$, which is a continuous function of t for $t > 0$ by Lemmas 10.1 and 10.2. If $t = \frac{1}{2}$, then (10.1) does not hold if $s = \frac{1}{2}$, which implies that $Q_{1/2,1/2}$ cannot be a cyclic quadrilateral. Thus $\min_O \left(\frac{1}{2}, \frac{1}{2}\right) > 0$, while $\min_I \left(\frac{1}{2}, \frac{1}{2}\right) = 0$ since $Q_{1/2,1/2}$ is a tangential quadrilateral. Hence $f\left(\frac{1}{2}\right) > 0$. Now $Q_{1/2,t_1}$ is a cyclic quadrilateral, which implies that $\min_O \left(\frac{1}{2}, t_1\right) = 0$, while $\min_I \left(\frac{1}{2}, t_1\right) > 0$ since $t_1 \neq \frac{1}{2}$ implies that $Q_{1/2,t_1}$ is not a tangential quadrilateral. Hence $f(t_1) < 0$, which implies that $f(t_2) = 0$ for some $t_2, \frac{1}{2} < t_2 < \frac{\sqrt{2}+1}{2}$ by the Intermediate Value Theorem. Since $\min_O \left(\frac{1}{2}, t_2\right) = \min_I \left(\frac{1}{2}, t_2\right), Q_{1/2,t_2}$ must be bielliptic. ∎

The proof above shows that $Q_{1/2,t_2}$ is bielliptic for some $t_2, \frac{1}{2} < t_2 < \frac{\sqrt{2}+1}{2}$. We now give an estimate for the value of t_2.

We want to estimate the solution of the equation $\min_I (s, t) \left(\frac{1}{2}, t\right) = \min_O \left(\frac{1}{2}, t\right)$. Equivalently, using the notation from the proofs of Lemmas 10.1 and 10.2 above, we want to estimate the solution of

$$G_t(r_I(t)) = Z_t(u_O(t)). \tag{10.2}$$

To simplify the notation in what follows, let $N = N_t(r_I(t)), O = O_t(r_I(t)), M = M_t(r_I(t)), u = u_O(t)$, and $H = H_t(u_O(t))$. Again, using the notation from the proofs of Lemmas 10.1 and 10.2, (10.2) is equivalent to

$$\frac{N}{(O + \sqrt{M})^2} = \frac{t^2(u+1)^2 - 4H}{(t(u+1) + 2\sqrt{H})^2} \tag{10.3}$$

subject to $p_t(r) = 0$. To simplify the notation further, we let $X = t^2(u+1)^2 - 4H$ and $Y = t^2(u+1)^2 + 4H$. Cross multiplying and squaring in (10.3) yields $N(Y + 4t(u+1)\sqrt{H}) = X(O^2 + M + 2O\sqrt{M}) \Rightarrow NY - X(O^2 + M) = -4t(u+1)N\sqrt{H} + 2XO\sqrt{M} \Rightarrow$
$(NY - X(O^2 + M))^2 = (-4t(u+1)N)^2 H + 4X^2O^2M - 16t(u+1) NXO\sqrt{H}\sqrt{M} \Rightarrow$

$$\left((NY - X(O^2 + M))^2 - (-4t(u+1)N)^2H - 4X^2O^2M\right)^2$$
$$- (-16t(u+1)NXO)^2 HM = 0. \tag{10.4}$$

The left hand side of (10.4) factors into

$$4096t^{10} \left(16t^4 - 32t^3 + 40t^2 - 8t + 1\right)^4 r^2 \cdot$$
$$\frac{(2t-1)^2 \, (r-1)^2 \, (r-2)^2 \, (r-2t)^2 \, q_t^2(r)}{\left(12t^2 - 4t + 1\right)^8},$$

where

$$
\begin{aligned}
q_t(r) \;=\; & \left(32t^6 - 80t^5 + 96t^4 - 72t^3 + 50t^2 - 9t + 1\right) r^4 + \\
& \left(-128t^6 + 224t^5 - 176t^4 + 80t^3 - 112t^2 + 22t - 3\right) r^3 + \\
& 2\left(96t^6 - 96t^5 + 64t^3 + 22t^2 - 6t + 1\right) r^2 \\
& -4t\left(32t^5 - 16t^4 - 16t^3 + 32t^2 - 6t + 1\right) r + \\
& t\left(2t - 1\right)\left(1 + 4t^2\right)^2.
\end{aligned}
$$

Of course this approach will yield many extraneous roots or values of t we don't want, such as $t = 0$ or $t = \dfrac{1}{2}$. In addition, $12t^2 - 4t + 1$ and $16t^4 - 32t^3 + 40t^2 - 8t + 1$ have no real roots. Using Maple, then, we entered the system of equations $p_t(r) = 0$, $q_t(r) = 0$, where

$$
\begin{aligned}
p_t(r) \;=\; & -\frac{1}{4}\left(2t+1\right)\left(2t-1\right)\left(2t-3\right) r^4 + 2t\left(4t^2 - 8t + 1\right) r^3 + \\
& \frac{1}{4}\left(-24t^3 + 84t^2 - 6t + 9\right) r^2 - \left(4t^3 + 8t^2 + 3t + 1\right) r + t\left(1 + 4t^2\right).
\end{aligned}
$$

One obtains a number of solutions, as expected, that do not work. However, one of the solutions gives t as the root of a polynomial, Z, of degree 20 and r as a rational function of t. One of the roots of Z which works is $t \approx 0.923\,79$, which then yields $r \approx 0.521\,68$; With these values $u_O(t) \approx 0.877\,98$, $\dfrac{N}{(O + \sqrt{M})^2} \approx$ 0.492 0, and $\dfrac{t^2(u+1)^2 - 4H}{(t(u+1) + 2\sqrt{H})^2} \approx 0.492\,0$; Thus with $t \approx 0.923\,79$, $Q_{1/2}, t$ is bielliptic of class $\alpha \approx \sqrt{1 - 0.492\,0} \sim 0.712\,7$; Using $u = 0.877\,98$ implies that E_O has approximate equation

$0.405\,53x^2 + 0.289\,90xy + 0.461\,90y^2 - 0.405\,53x - 0.461\,90y = 0$; E_I has approximate equation $49.38x^2 + 6.850\,3xy + 25y^2 - 36.659x - 26.084y = -6.8038$.

Remark 10.1 *An alternate approach, similar to what we did in [16], is to consider the family of quadrilaterals $\{Q_\lambda\} \subset \{Q_{s,t}\}$ given by $s(\lambda) = -\dfrac{1}{4}\lambda + \dfrac{3}{4}$, $t(\lambda) = \left(\dfrac{1}{\sqrt{2}} - \dfrac{1}{4}\right)\lambda + \dfrac{3}{4}, 0 \le \lambda \le 1$; $\lambda = 0$ gives $s = t = \dfrac{3}{4}$ and thus Q_0 is a tangential quadrilateral which is not cyclic; $\lambda = 1$ gives $s = \dfrac{1}{2}$ and*

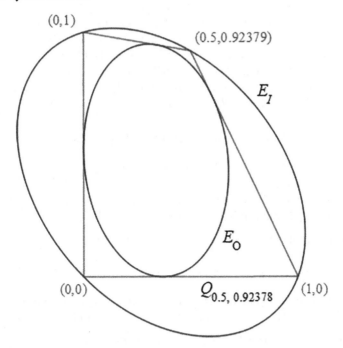

FIGURE 10.1
Bielliptic quadrilateral $Q_{0.5,0.92379}$

$t = \frac{1}{2}\sqrt{2} + \frac{1}{2}$ *and hence* Q_1 *is a cyclic quadrilateral which is not tangential. Then the Intermediate Value Theorem would apply as above. One would have to show continuity as done in Lemmas 10.1 and 10.2.*

For *parallelograms*, we have the following result:

Theorem 10.2 *Let Q be a parallelogram, let p and q denote the lengths of the diagonals of Q, and let s_1 and s_2 denote the lengths of the intersecting sides of Q(we do not assume any ordering here). Then Q is bielliptic if and only if one of the following holds:*
(i) $p^2 = 2s_1^2$ and $q^2 = 2s_2^2$ or (ii) $p^2 = 2s_2^2$ and $q^2 = 2s_1^2$

Proof. Suppose first that Q is tangential. If Q is bielliptic, then Q must be cyclic and thus Q is bicentric. A bicentric parallelogram must be a square, and (i) and (ii) both hold in that case. Conversely, suppose that (i) holds. Corrollary 2.2 implies that Q is a rhombus, which implies that $s_1 = s_2$. Thus $p = q$, which implies that Q is a square and is thus cyclic(and hence bicentric). The proof is similar if (ii) holds. Assume now that Q is **not tangential**. Now suppose that Q is cyclic. Then the necessary part of Theorem 10.2 cannot hold since Q cannot be bielliptic. Corollary 5 implies that Q is a rectangle. If (i) holds or (ii) holds, then Q must be a square. But that contradicts the assumption that Q is not tangential. Thus the sufficiency part of Theorem 10.2 cannot hold either. So assume also that Q is **not cyclic**. By Lemma 2.4 we may assume that $Q = Q_{d,k,l}$ for some $l, k > 0;\ 0 \le d < l$, where $Q_{d,k,l}$ is neither cyclic nor tangential. This will allow us to apply formulas from the proofs of Theorem 2.3 and 9.1. Recalling the notation in (2.6), one has $p^2 = 4(J + 2ld), q^2 = 4(J - 2ld), s_1^2 = 4(d^2 + k^2), s_2^2 = 4l^2$. Also,

$$\begin{aligned} p^2 - 2s_1^2 &= -4(N - 2ld), q^2 - 2s_2^2 = 4(N - 2ld), \\ p^2 - 2s_2^2 &= 4(N + 2ld), q^2 - 2s_1^2 = -4(N + 2ld). \end{aligned} \tag{10.5}$$

Let E_1 denote an ellipse inscribed in $Q_{d,k,l}$ and let a and b denote the lengths of the semi–major and semi–minor axes, respectively, of E_1.

By the proof of Theorem 2.3(see (2.8) and (2.9)), $\dfrac{b^2}{a^2} = f(u) = \dfrac{4k^2l^2\,(1 - u^2)}{(g(u) + \sqrt{q(u)})^2}, -1 < u < 1$, where $g(u) = J + 2dlu$ and $q(u) = (2uld + I)^2 + 4k^2\,(ul + d)^2$(see (2.7), with $g(u)$ replacing $p(u)$ to avoid confusion with p above). Again by the proof of Theorem 2.3, the unique ellipse, E_I, of minimal eccentricity inscribed in $Q_{d,k,l}$ corresponds to $u = u_\epsilon = -\dfrac{2dl}{J}$; Now

$g(u_\epsilon) = J + 2dl\left(-\dfrac{2dl}{J}\right) = \dfrac{J^2 - 4l^2d^2}{J}$ and $q(u_\epsilon) = \left(2\left(-\dfrac{2dl}{J}\right)ld + I\right)^2 + 4k^2\left(l\left(-\dfrac{2dl}{J}\right) + d\right)^2 = \dfrac{(J^2 - 4l^2d^2)N^2}{J^2}$. Note that $J^2 - 4l^2d^2 = ((d+l)^2 + k^2)((d-l)^2 + k^2) > 0$. Hence $g(u_\epsilon) + \sqrt{q(u_\epsilon)} = \dfrac{J^2 - 4l^2d^2}{J} + \dfrac{\sqrt{J^2 - 4l^2d^2}\,|N|}{J} = \dfrac{\sqrt{J^2 - 4l^2d^2}}{J}(\sqrt{J^2 - 4l^2d^2} + |N|)$ and $1 - u_\epsilon^2 = 1 - \left(\dfrac{2dl}{J}\right)^2 = \dfrac{J^2 - 4l^2d^2}{J^2}$; Simplifying yields

$$f(u_\epsilon) = \frac{4k^2l^2}{(\sqrt{J^2 - 4l^2d^2} + |N|)^2}. \tag{10.6}$$

Now let E_2 denote an ellipse circumscribed about $Q_{d,k,l}$ and let a and b denote the lengths of the semi–major and semi–minor axes, respectively, of E_2. By the proof of Theorem 9.1, $\dfrac{b^2}{a^2} = Z(u)$, where $Z(u)$ and $H(u)$ are given by (9.3) and the unique ellipse, E_O, of minimal eccentricity circumscribed about $Q_{d,k,l}$ corresponds to $u = u_0 = \dfrac{Jk^2}{k^2 + 2d^2}$; Simplifying yields $H(u_0) = 4d^2k^2\dfrac{(l^2 - d^2)^2(d^2 + k^2)}{(2d^2 + k^2)^2} \Rightarrow \sqrt{H(u_0)} = 2dk\dfrac{(l^2 - d^2)\sqrt{d^2 + k^2}}{2d^2 + k^2}$; $u_0 - N = 2\dfrac{(l^2 - d^2)(d^2 + k^2)}{2d^2 + k^2}$ and some more simplification gives

$$Z(u_0) = \frac{k^2}{\left(\sqrt{d^2 + k^2} + d\right)^2}. \tag{10.7}$$

By (10.6) and (10.7), the eccentricities of E_I and E_O are equal if and only if

$$\frac{4k^2l^2}{(\sqrt{J^2 - 4l^2d^2} + |N|)^2} = \frac{k^2}{\left(\sqrt{d^2 + k^2} + d\right)^2} \Longleftrightarrow$$

$$2l\left(\sqrt{d^2 + k^2} + d\right) = \sqrt{J^2 - 4l^2d^2} + |N|. \tag{10.8}$$

Note that if $N = 0$, then $d^2 + k^2 = l^2$, which implies that $Q_{d,k,l}$ is a rhombus. Then by Lemma 2.2, $Q_{d,k,l}$ is tangential. Since we have assumed that $Q_{d,k,l}$ is not tangential, we may assume that $N \neq 0$. Assume first that $N > 0$. Then (10.8) holds $\Longleftrightarrow 2l\left(\sqrt{d^2 + k^2} + d\right) - N = \sqrt{J^2 - 4l^2d^2} \Longleftrightarrow 4l^2\left(\sqrt{d^2 + k^2} + d\right)^2 - 4lN\left(\sqrt{d^2 + k^2} + d\right) + N^2 = J^2 - 4l^2d^2$ (note that the left hand side is positive since the right hand side is positive, so the implication goes both ways) $\Longleftrightarrow (8l^2d - 4lN)\sqrt{d^2 + k^2} = J^2 - 4l^2d^2 - 8l^2d^2 - 4l^2k^2 + 4lNd - N^2 \Longleftrightarrow l(2ld - N)\sqrt{d^2 + k^2} = ld(N - 2ld) \Longleftrightarrow N - 2ld = 0$ (since $\sqrt{d^2 + k^2} + d \neq 0$) $\Longleftrightarrow$
(i) holds by (10.5). Similarly, if $N < 0$, then one can show that (10.8) holds $\Longleftrightarrow N + 2ld = 0 \Longleftrightarrow$ (ii) holds by (10.5). ∎

Theorem 10.3 *There exists a bielliptic parallelogram of class α for each $0 < \alpha < 1$.*

Proof. Working with $Q_{d,k,l}$, by the proof of Theorem 10.2, $Q_{d,k,l}$ is bielliptic, if and only if $N = 2dl$ or $N = -2dl$. It turns out that $N = -2dl$ is the choice that works here. Given $0 < \alpha < 1$, let $\tau = 1 - \alpha^2$. The unique ellipse, E_O, of minimal eccentricity circumscribed about $Q_{d,k,l}$ has eccentricity $\alpha \Longleftrightarrow Z(u_0) = \tau$. By (10.7), $Z(u_0) = \tau \Rightarrow k^2 = \tau(d^2 + k^2) + 2\tau d\sqrt{d^2 + k^2} + \tau d^2 \Rightarrow 2\tau d\sqrt{d^2 + k^2} = (1 - \tau)k^2 - 2\tau d^2 \Rightarrow$
$$4\tau^2 d^2(d^2 + k^2) = (1 - \tau)^2 k^4 - 4\tau(1 - \tau)d^2k^2 + 4\tau^2 d^4 \Rightarrow k^2((1 - \tau)^2 k^2 - 4\tau d^2) = 0 \Rightarrow k^2 = \frac{4\tau}{(1 - \tau)^2}d^2 \Rightarrow k = \frac{2\sqrt{\tau}}{1 - \tau}d$ \text{ and } d^2 + k^2 = \left(1 + \frac{4\tau}{(1 - \tau)^2}\right)d^2 = \frac{(1 + \tau)^2}{(1 - \tau)^2}d^2; \ N = -2ld \Rightarrow d^2 + k^2 - l^2 = -2ld \Rightarrow l^2 - 2dl - \frac{(1 + \tau)^2}{(1 - \tau)^2}d^2 = 0;$$ solving for l yields $l = \frac{(1 - \tau) \pm \sqrt{2}\sqrt{1 + \tau^2}}{1 - \tau}d = \left(1 \pm \frac{\sqrt{2}\sqrt{1 + \tau^2}}{1 - \tau}\right)$; Since we require $l > 0$ and
$\frac{l}{d} > 1$, $l = \left(1 + \frac{\sqrt{2}\sqrt{1 + \tau^2}}{1 - \tau}\right)d$ is the correct choice. With this choice,

$d^2 + k^2 - l^2 + 2ld = 0$, which implies that $N = -2dl$. Hence $Q_{d,k,l}$ is bielliptic by (10.5) and by Theorem 10.2(ii). ∎

Summarizing: Given $0 < \tau < 1$ and $d > 0$, let

$$k = \frac{2\sqrt{\tau}}{1 - \tau} d \text{ and} \tag{10.9}$$

$$l = \left(1 + \frac{\sqrt{2}\sqrt{1 + \tau^2}}{1 - \tau}\right) d.$$

Then $Q_{d,k,l}$ is a bielliptic parallelogram of class $\alpha = \sqrt{1 - \tau}$.

Examples: (1) Consider $Q_{d,k,l}$ with $d = 2$ and suppose that we want a bielliptic parallelogram of class $\alpha = \dfrac{1}{\sqrt{2}}$, which implies that $\tau = \dfrac{1}{2}$. By (10.9) one has $k = 4\sqrt{2}$ and $l = 2(1 + \sqrt{10})$. One can verify that $N = -8(1 + \sqrt{2}\sqrt{5}) < 0$, $f(u_\epsilon) = \dfrac{4k^2 l^2}{(\sqrt{J^2 - 4l^2 d^2} + |N|)^2} = \dfrac{1}{2}$, and $Z(u_0) = \dfrac{k^2}{(\sqrt{d^2 + k^2} + d)^2} = \dfrac{1}{2}$.
The common eccentricity of E_I and E_O is $\dfrac{1}{\sqrt{2}}$.

(2) Consider the parallelogram, Q_1, with vertices $(1, 2), (3, 6), (1 + \sqrt{39}, 3)$, and $(-1 + \sqrt{39}, -1)$. The squares of the lengths of the diagonals are $p^2 = 40$ and $q^2 = 8(13 - \sqrt{39})$ and the squares of the lengths of the sides are $s_1^2 = 20$ and $s_2^2 = 4(13 - \sqrt{39})$; Then $p^2 = 2s_1^2$ and $q^2 = 2s_2^2$, which implies that Q_1 is bielliptic by Theorem 10.2. Let E_I and E_O denote the inscribed and circumscribed ellipses of minimal eccentricity, respectively, for Q_1, let $\min_I =$ eccentricity of E_I, and let $\min_O =$ eccentricity of E_O.

(a) Find the common value of $\min_I$ and $\min_O$. We find an isometry, $\mathcal{F}$, such that $\mathcal{F}(Q_1) = Q_{d,k,l}$ for some $l, k > 0, 0 \leq d < l$. First, translate the center of Q_1, $\left(1 + \dfrac{\sqrt{39}}{2}, \dfrac{5}{2}\right)$, to $(0, 0)$. That yields the parallelogram, Q_2, with vertices $-\dfrac{1}{2}(\sqrt{39}, 1), \dfrac{1}{2}(4 - \sqrt{39}, 7), \dfrac{1}{2}(\sqrt{39}, 1)$, and $-\dfrac{1}{2}(4 - \sqrt{39}, 7)$. Second, rotate Q_2 so that there are two parallel horizontal sides. That is done by rotating counterclockwise about the origin thru the angle, θ, where $\cos\theta = \dfrac{1}{m}(\sqrt{39} - 2), \sin\theta = \dfrac{3}{m}$, and

$$m = 2\sqrt{13 - \sqrt{39}}.$$

The corresponding rotation matrix is

$$M = \frac{1}{m}\begin{bmatrix} \sqrt{39} - 2 & -3 \\ 3 & \sqrt{39} - 2 \end{bmatrix},$$

and multiplying each vertex of Q_2 by M yields the parallelogram, Q_3, with vertices $\frac{1}{m}(\sqrt{39}-18, 1-2\sqrt{39})$, $\frac{1}{m}(-34+3\sqrt{39}, 2\sqrt{39}-1)$, $-\frac{1}{m}(\sqrt{39}-18, 1-2\sqrt{39})$, $-\frac{1}{m}(-34+3\sqrt{39}, 2\sqrt{39}-1)$. Finally, reflect Q_3 thru the x axis to obtain the parallelogram, Q_4, with vertices $\frac{1}{m}(-34+3\sqrt{39}, 1-2\sqrt{39})$, $\frac{1}{m}(\sqrt{39}-18, 2\sqrt{39}-1)$, $-\frac{1}{m}(-34+3\sqrt{39}, 1-2\sqrt{39})$, $-\frac{1}{m}(\sqrt{39}-18, 2\sqrt{39}-1)$. It is then easy to show that Q_4 equals the parallelogram, $Q_{d,k,l}$, where

$$k = \frac{1}{m}(2\sqrt{39}-1),$$

$$l = \frac{m}{2},$$

$$d = \frac{1}{m}(8-\sqrt{39}).$$

If a and b denote the lengths of the semi–major and semi–minor axes, respectively, of E_O, then by the proof of Theorem 10.2, $\dfrac{b^2}{a^2} = \dfrac{k^2}{\left(\sqrt{d^2+k^2}+d\right)^2}$,

which implies that $\min_I = \min_O = \sqrt{1-\dfrac{b^2}{a^2}} = \dfrac{\sqrt{\left(\sqrt{d^2+k^2}+d\right)^2 - k^2}}{\sqrt{d^2+k^2}+d} \approx$

0.512; Note that Q_1 and Q_4 are isometric, so the values of a and b do not change.

(b) Find the equations of, and plot, E_I and E_O along with the parallelogram, Q_1. As noted above, by the proof of Theorem 2.3, one obtains $\mathcal{F}(E_1)$ by letting $u = -\dfrac{2dl}{d^2+k^2+l^2} = \dfrac{143-21\sqrt{39}}{31\sqrt{39}-273} \approx -0.149\,30$ in (2.4). That yields the approximate equation

$$4.886x^2 + 0.222\,84xy + 6.607\,0y^2 = 32.269. \tag{10.10}$$

Similarly, by the proof of Theorem 9.1, one obtains $\mathcal{F}(E_O)$ by letting $u = \dfrac{k^2(d^2+k^2+l^2)}{k^2+2d^2} = \dfrac{1}{3}\dfrac{47\,697-5959\sqrt{39}}{2041-277\sqrt{39}} \approx 11.231$ in (9.1). That yields the approximate equation

$$14.025x^2 - 4.284\,4xy + 14.680y^2 = 164.86. \tag{10.11}$$

Now $M \approx \begin{bmatrix} 0.816\,65 & -0.577\,14 \\ 0.577\,14 & 0.816\,65 \end{bmatrix}$ and the center of Q_1 is approximately $(4.122\,5, 2.5)$. Thus, to obtain the corresponding approximate equations for E_I and E_O, do the following: First replace y by $-y$; then replace x by $0.816\,65x - 0.577\,14y$ and replace y by $0.577\,14x + 0.816\,65y$; Finally, replace x by $x - 4.122\,5$ and replace y by $y - 2.5$ in both (10.10) and in (10.11). After simplifying, that gives $5.354\,3x^2 + 1.547\,9xy + 6.138\,8y^2 - 48.016x - 37.075y = -113.05$ as the approximate equation of E_I, and

$$16.263x^2 + 2.047\,7xy + 12.443y^2 - 139.2x - 70.655y = -210.39$$ as the approximate equation of E_O.

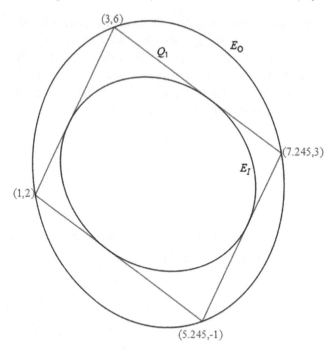

FIGURE 10.2
Bielliptic parallelogram with vertices $(1, 2), (3, 6), (1 + \sqrt{39}, 3), (1 + \sqrt{39}, 1)$

Remark 10.2 *We have demonstrated a convex quadrilateral, Q, which is not a parallelogram and which is bielliptic of class α for some $0 < \alpha < 1$. We have also shown that there exists a bielliptic parallelogram of class α for each $0 < \alpha < 1$. It is natural to ask about bielliptic trapezoids which are not parallelograms. In ([16], Theorem 5) we stated that there is a bielliptic trapezoid, Q, which is not a parallelogram and which is bielliptic of class α for some $0 < \alpha < 1$. However, there is an error in the proof. Let $Q_{1,t}$ denote $Q_{s,t}$ with $s = 1$. One can show that $Q_{1,t}$ is not bielliptic when $t \neq 1$. This leaves open the possibility that the trapezoid $Q_{s,t,s,w}$ is bielliptic for some s, t, w.*

11

Algorithms for Circumscribed Ellipses

11.1 Minimal Area and Minimal Eccentricity for Non-Parallelograms

Here we use material from Proposition 8.1 and from the proof of Theorem 8.3.

(1) To find the unique ellipse of minimal area, E_A, circumscribed about a convex quadrilateral, Q, which is **not** a parallelogram:

   (i) Find a nonsingular affine transformation, $\mathbb{T}_{s,t}$, so that $\mathbb{T}_{s,t}(Q) = Q_{s,t}$ for some $(s,t) \in G$–see (1) in § 7/Transformations.

   (ii) Let u_1 and u_2 equal $t \dfrac{st + s + t - 1 \pm 2\sqrt{st}\sqrt{s + t - 1}}{s(s-1)^2}$, with u_1 equal to the negative square root. Find the unique root, $u_0 \in (u_1, u_2)$, of the cubic polynomial

   $$q(u) = s^3 (s-1)^2 u^3 + s^2 t(2(s-1)^2 + st + s + t - 1)u^2$$
   $$- st^2(2(t-1)^2 + st + s + t - 1)u - t^3(t-1)^2 .$$

   (iii) The equation of the ellipse, $E_{s,t}$, of minimal area circumscribed about $Q_{s,t}$ is then given by $stu_0 x^2 + (t(1-t) + s(1-s)u_0)xy + sty^2 - stu_0 x - sty = 0$.

   (iv) One then has $E_A = \mathbb{T}_{s,t}^{-1}(E_{s,t})$.

(2) To find the unique ellipse of minimal eccentricity, E_O, circumscribed about a convex quadrilateral, Q, which is **not** a parallelogram:

   (i) Find an affine transformation, $\mathbb{T}_{s,t,v,w}$, which preserves the eccentricity of ellipses and so that $\mathbb{T}_{s,t,v,w}(Q) = Q_{s,t,v,w}$, where (1.10), (1.12), and (1.13) hold–see (2) in § 7/Transformations.

   (ii) Let $f_1 = v(t-1) + (1-w)s$, $f_2 = vt - ws$, $f_3 = sv(t-w)$, and $f_4 = sv(s-v)$. Let $u_0 = \dfrac{(sw^2 - t^2 v + f_2)^2 + f_3^2 + svf_1f_2}{f_3^2 + f_4^2 + svf_1f_2}$.

DOI: 10.1201/9781003474890-14

Then the equation of the unique ellipse of minimal eccentricity, $E_{s,t,v,w}$, circumscribed about $Q_{s,t,v,w}$ is given by $f_3 u_0 x^2 + (sw^2 - t^2 v + f_2 - f_4 u_0)xy + f_3 y^2 + (f_1 tw - f_2 svu_0)x - f_3 y = 0$.

(iii) One then has $E_I = \mathbb{F}^{-1}_{s,t,v,w}(E_{s,t,v,w})$.

11.2 Minimal Area and Minimal Eccentricity for Parallelograms

(1) To find the unique ellipse of minimal area, E_A, circumscribed about a parallelogram, Q:

   (i) Let S be the square with vertices $(-1,-1), (-1,1), (1,1)$, and $(1,-1)$. Find a nonsingular affine transformation, $\mathbb{F}$, so that $\mathbb{F}(Q) = S$.

   (ii) The ellipse, $E_{s,t}$, of minimal area circumscribed about S is the circle $x^2 + y^2 = 2$.

   (iii) One then has $E_A = \mathbb{F}^{-1}_{s,t}(E_{s,t})$.

(2) To find the unique ellipse of minimal eccentricity, E_O, circumscribed about a parallelogram, Q:

Let $Q_{d,k,l}$ be the parallelogram with vertices $(-l - d, -k), (-l + d, k), (l + d, k)$, and $(l - d, -k)$, where $l, k > 0, d \geq 0, d < l$.

   (i) Find an isometry, $\mathbb{F}$, so that $\mathbb{F}(Q) = Q_{d,k,l}$.

   (ii) Let $u_0 = k^2 \dfrac{l^2 + k^2 + d^2}{k^2 + 2d^2}$. The equation of the unique ellipse of minimal eccentricity, E_P, circumscribed about $Q_{d,k,l}$ is then given by $k(u_0 - k^2)x^2 - 2d(u_0 - k^2)xy + k\left(l^2 - d^2\right)y^2 - k\left(l^2 - d^2\right)u_0 = 0$.

   (iii) One then has $E_I = \mathbb{F}^{-1}(E_P)$.

12

Related Research and Open Questions

12.1 Arc Length

Proving results related to maximizing or minimizing the arc length of ellipses inscribed in, or circumscribed about, convex quadrilaterals appears, as one might expect, to be a bit more difficult than for maximizing or minimizing the area or eccentricity of ellipses inscribed in, or circumscribed about, convex quadrilaterals. Even for parallelograms the arc length problem is non-trivial, though it seems fairly doable using elliptic functions. The only result we have proven so far was in Theorem 2.4 in § 2. We proved that there is a unique ellipse of maximal arc length inscribed in any rectangle, Z. We do not yet have a proof for circumscribed ellipses–in that case one would try to show that there is a unique ellipse of minimal arc length circumscribed about a given rectangle. For ellipses inscribed in a rectangle, Z, we also showed in Theorem 2.4 that the unique ellipse of maximal area inscribed in Z, the unique ellipse of minimal eccentricity inscribed in Z, and the unique ellipse of maximal arc length inscribed in Z are all the same ellipse. Does that also hold for circumscribed ellipses?

12.2 Bielliptic Quadrilaterals

There are many known properties and equivalences for bicentric quadrilaterals. We have found an equivalence for bielliptic parallelograms, but it would be interesting to find some properties of bielliptic quadrilaterals in general. It might also be interesting to investigate the class of bielliptic midpoint diagonal quadrilaterals.

DOI: 10.1201/9781003474890-15

12.3   Other Families of Curves

Let F be a family of simple closed curves in the xy plane which is invariant under isometries and let Q be a convex quadrilateral. Examples of such families might be what are often referred to as oval or egg curves, such as $\left|\dfrac{x}{a}\right|^{2.5} + \left|\dfrac{y}{b}\right|^{2.5} = 1$ or $\dfrac{x^2}{a} + \dfrac{y^2}{b}(1+0.2x) = 1$. Of course one would apply suitable transformations to these curves to make them invariant under isometries. Below is a plot of the curve $x^4 + y^4 = 1$ inscribed in the square with vertices $(\pm 1, \pm 1)$. The points of tangency are $(-1,0), (0,1), (1,0), (0,-1)$.

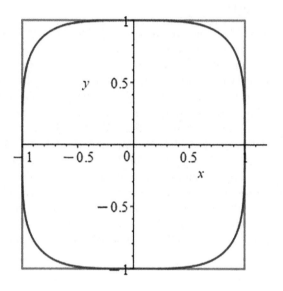

FIGURE 12.1
The curve $x^4 + y^4 = 1$ inscribed in a square

Natural questions to ask are: Given a convex quadrilateral, Q,

(i)   What is the locus of centers of curves in F inscribed in Q?

(ii) Does there always exist a unique curve in F of maximal area inscribed in Q or a unique curve in F of minimal area circumscribed about Q? One can sk similar questions involving a suitable generalization of the eccentricity for ellipses.

It would be useful to obtain equations for such curves inscribed in, or circumscribed about, Q as we did for ellipses in Proposition 1.3 and in Proposition 8.1 for non-parallelograms. Of course one could also look at curves in F which are inscribed in, or circumscribed about parallelograms.

(iii) In §6 we showed that if P is a point in the interior of Q other than the intersection point of the diagonals, then there is at least one ellipse passing through P and inscribed in Q. Does this still hold for curves in F?

A

Appendix

A.1 General Results on Ellipses

Throughout we let $G(x,y) = Ax^2 + Bxy + Cy^2 + Dx + Ey + F$, where all of the coefficients are real and where $A, C > 0$.

A.1.1 Coefficient formulas

Given a conic, E_0, with equation $G(x,y) = 0$, we define the following:

$$\begin{aligned} \Delta &= 4AC - B^2, \\ \delta &= CD^2 + AE^2 - BDE - F\Delta. \end{aligned} \tag{A.1}$$

If E_0 is an ellipse, we let a and b denote the lengths of the semi–major and semi–minor axes, respectively, of E_0 and we let (x_0, y_0) denote the center of E_0. We state the following useful general lemma about when the coefficients of G define an ellipse. The content of the following lemma is well known and can be found, for example, in [41]. Also see [24]. The notation we use here is slightly different. We omit any proof.

Lemma A.1 *Let Γ be the conic with equation $G(x,y) = 0$. Then Γ is an ellipse if and only if $\Delta > 0$ and $\delta > 0$.*

Note that there is no loss of generality in assuming that $A, C > 0$. If A and C have opposite signs, then $G(x,y) = 0$ cannot give an ellipse. If $A, C < 0$, then one can multiply thru in the equation by -1 so that $A, C > 0$.

The following lemma is key and is cited extensively in many of the sections of this book.

Lemma A.2 *Suppose that E_0 is an ellipse with equation $G(x,y) = 0$ and let Δ and δ be as in A.1. Then*

$$a^2 b^2 = \frac{4\delta^2}{\Delta^3}, \tag{A.2}$$

$$\frac{b^2}{a^2} = \frac{A + C - \sqrt{(A-C)^2 + B^2}}{A + C + \sqrt{(A-C)^2 + B^2}}, \tag{A.3}$$

and E_0 has center (x_0, y_0), where

$$x_0 = \frac{BE - 2CD}{\Delta}, y_0 = \frac{BD - 2AE}{\Delta}. \qquad (A.4)$$

Proof. By [42],

$$
\begin{aligned}
a^2 &= \mu \frac{A + C + \sqrt{(A-C)^2 + B^2}}{2} \\
b^2 &= \mu \frac{A + C - \sqrt{(A-C)^2 + B^2}}{2}
\end{aligned}
\qquad (A.5)
$$

where $\mu = \dfrac{4\delta}{\Delta^2}$. By (A.5), $a^2 b^2 = \dfrac{\mu^2}{4}((A+C)^2 - (A-C)^2 - B^2) = \left(\dfrac{4\delta^2}{\Delta^4}\right)\Delta = \dfrac{4\delta^2}{\Delta^3}$. (A.3) follows immediately from (A.5). (A.4) can be found in [42]. ∎

Since ellipses, tangent lines to ellipses, ratios of areas of ellipses, and four-sided convex polygons are preserved under nonsingular affine transformations ([43]), we have the two following lemmas.

Lemma A.3 *Suppose that E_0 is an ellipse inscribed in the convex quadrilateral, Q, and let $\mathcal{F}$ be a non–singular affine transformation. Then $\mathcal{F}(E_0)$ is inscribed in $\mathcal{F}(Q)$. Furthermore, if E_0 is tangent to Q at ζ_1–ζ_4, then $\mathcal{F}(E_0)$ is tangent to $\mathcal{F}(Q)$ at $\mathcal{F}(\zeta_1)$–$\mathcal{F}(\zeta_4)$.*

Lemma A.4 *Suppose that E_0 is the unique ellipse of maximal area inscribed in the convex quadrilateral, Q, and let $\mathcal{F}$ be a nonsingular affine transformation. Then $\mathcal{F}(E_0)$ is the unique ellipse of maximal area inscribed in $\mathcal{F}(Q)$.*

In addition to scaling transformations, the following lemma concerns what happens to the eccentricity of an ellipse under an isometry. We omit the proof.

Lemma A.5 *Suppose that E_0 is the unique ellipse of minimal eccentricity inscribed in the convex quadrilateral, Q, and let $\mathcal{F}$ be an isometry. Then $\mathcal{F}(E_0)$ is the unique ellipse of minimal eccentricity inscribed in $\mathcal{F}(Q)$.*

The following result expresses the foci of an ellipse, E_0, as a function of the coefficients of an equation of E_0 and of the length of the major axis of E_0. Most of the details can be found in [44]. We use this result in § 2.5 to prove part of Theorem 2.6.

Lemma A.6 *Let E_0 be an ellipse which is not a circle. Let θ denote the counterclockwise angle of rotation to the major axis of E_0 from the line thru the center of E_0 and parallel to the x axis, with $0 \le \theta < \pi$. Let $a =$ length of semi–major and $b =$ length of semi–minor axes of E_0, respectively. Let F_1 and F_2 denote the foci of E_0, with $F_2 = (x_c, y_c)$ the rightmost focus (if $\theta = \dfrac{\pi}{2}$, we* *let F_2 denote the uppermost focus) and let (x_0, y_0) be the center of E_0. Then E_0 can be written in the form $A(x - x_0)^2 + B(x - x_0)(y - y_0) + C(y - y_0)^2 - a^2 b^2 = 0$, where $A, C > 0$.*

(i) If $B \neq 0$, then the foci of E_0 are given by

$$\begin{aligned} F_1 &= (x_0 - \sqrt{a^2 - A}, y_0 + (\operatorname{sgn} B)\sqrt{a^2 - C}), \\ F_2 &= (x_0 + \sqrt{a^2 - A}, y_0 - (\operatorname{sgn} B)\sqrt{a^2 - C}). \end{aligned} \qquad (A.6)$$

In addition, if $0 \leq \theta < \dfrac{\pi}{2}$, then $B < 0$, while if $\dfrac{\pi}{2} < \theta < \pi$, then $B > 0$.

(ii) let $r = 1 - \operatorname{sgn}(A - C), s = 1 + \operatorname{sgn}(A - C)$. If $B = 0$, then $\theta = \dfrac{\pi}{2}$ and the foci of E_0 are given by

$$\begin{aligned} F_1 &= (x_0 - \frac{r}{2}\sqrt{a^2 - A}, y_0 - \frac{s}{2}\sqrt{a^2 - C}), \\ F_2 &= (x_0 + \frac{r}{2}\sqrt{a^2 - A}, y_0 + \frac{s}{2}\sqrt{a^2 - C}). \end{aligned} \qquad (A.7)$$

A.1.2   Conjugate diameters

Given a diameter, l, of an ellipse, E_0, there is a unique diameter, m, of E_0 such that the midpoints of all chords parallel to l lie on m. In this case we say that l and m are conjugate diameters of E_0, or that m is a diameter of E_0 conjugate to l.

In the figure below, PP' and QQ' are conjugate diameters. See https://commons.wikimedia.org/wiki/File:Ellipse_conjugated_diameter.svg and https://creativecommons.org/licenses/by/4.0/deed.en.

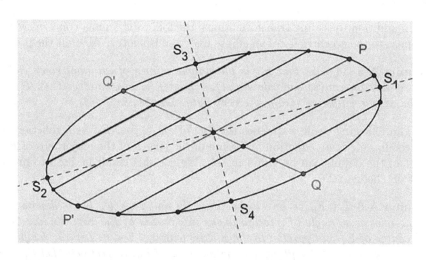

FIGURE A.1
Illustrating the definition of conjugate diameters of an ellipse

A pair of **perpendicular** conjugate diameters of an ellipse are the axes of that ellipse. Here are some useful facts and some terminology about conjugate diameters.

1. Every ellipse which is not a circle has a unique pair of **equal** conjugate diameters–that is, where $|l| = |m|$.

2. Let θ_1 and θ_2 be the angles which a pair of conjugate diameters make with the positive x axis. Then $\tan \theta_1$ and $\tan \theta_2$ are called a pair of **conjugate directions** or **directional constants**.

The following lemma shows that affine transformations send conjugate diameters to conjugate diameters. The proof follows immediately from the properties of affine transformations.

Lemma A.7 *Let $\mathcal{F}: R^2 \to R^2$ be an affine transformation and suppose that l and m are conjugate diameters of an ellipse, E_0. Then $l' = \mathcal{F}(l)$ and $m' = \mathcal{F}(m)$ are conjugate diameters of $E_0' = \mathcal{F}(E_0)$.*

The following lemma shows that the scaling transformations preserve the eccentricity of ellipses, as well as the property of the equal conjugate diameters of an ellipse being parallel to the diagonals of Q. The scaling transformations are a specific class of affine transformations which are **not** necessarily isometries.

Lemma A.8 *Let $\mathcal{F}$ be the scaling transformation given by $\mathcal{F}(x, y) = (kx, ky), k \neq 0$ and let E_0 be an ellipse.*

(i) Then $\mathcal{F}(E_0)$ is also an ellipse.

(ii) E_0 and $\mathcal{F}(E_0)$ have the same eccentricity for any ellipse, E_0.

(iii) If the equal conjugate diameters of E_0 are parallel to the diagonals of a quadrilateral, Q, then the equal conjugate diameters of $\mathcal{F}(E_0)$ are parallel to the diagonals of $\mathcal{F}(Q)$.

Proof. (i) follows immediately from Lemma A.1. (ii) follows easily and we omit the proof. To prove (iii), suppose that l and m are equal conjugate diameters of an ellipse, E_0, which are parallel to the diagonals of Q. By Lemma A.7, $l' = \mathcal{F}(l)$ and $m' = \mathcal{F}(m)$ are conjugate diameters of $E_0' = \mathcal{F}(E_0)$. Let (x_1, y_1) and (x_2, y_2) be the points of intersection of l with E_0, and let (x_3, y_3) and (x_4, y_4) be the points of intersection of m with E_0. Then $|(x_1, y_1)\,(x_2, y_2)| = |(x_3, y_3)\,(x_4, y_4)|$, which implies that $|\mathcal{F}(x_1, y_1)\,\mathcal{F}(x_2, y_2)| = k\,|(x_1, y_1)\,(x_2, y_2)| = k\,|(x_3, y_3)\,(x_4, y_4)| = |\mathcal{F}(x_3, y_3)\,\mathcal{F}(x_4, y_4)|$ and hence l' and m' are equal conjugate diameters of $\mathcal{F}(E_0)$. Since affine transformations take parallel lines to parallel lines, l' and m' are parallel to the diagonals of $\mathcal{F}(Q)$. ∎

A.2   Proofs of Some Earlier Results

A.2.1   Proposition 1.1

We prove (ii) first. The following two lemmas are key:

Lemma A.9 *If the tangents to an ellipse at two points A and B meet at a point P, then the line joining P to the center O of the ellipse bisects the chord AB.*

Lemma A.9 alone is sufficient to prove Proposition 1.1, but using the following lemma makes the proof somewhat shorter.

Lemma A.10: *Suppose that E_0 is an ellipse tangent to the quadrilateral, $Q = Q(A_1, A_2, A_3, A_4)$ (see 1), at the points $\zeta_1 - \zeta_4$, where $\zeta_j \in S_j, j = 1, ..., 4$. Then $\overleftrightarrow{\zeta_1 \zeta_3}$ and $\overleftrightarrow{\zeta_2 \zeta_4}$ intersect at the intersection point of the diagonals of Q.*

Proof. The proof of Lemma A.10 for circles inscribed in tangential quadrilaterals can be found in various places and we omit the details. See, for example, Notes on Euclidean Geometry by Paul Yiu, pages 156-157. Lemma A.10 for ellipses then follows easily using affine invariance.

We will use Lemmas A.9 and A.10 to derive formulas for the points of tangency and the center in terms of s, t, and x_4. That will then allow us to find the equation. Recall that the lines, going clockwise, which make up

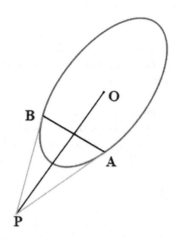

FIGURE A.2
Illustrating Lemma A.9

the boundary of $Q_{s,t}$ are given by L_1: $x = 0, L_2$: $y = 1 + \dfrac{t-1}{s}x, L_3$: $y = \dfrac{t}{s-1}(x-1)$, and L_4: $y = 0$. Recall that the center of E_0 is (h_0, L_0), where

$$h_0 \in I = \begin{cases} \left(\dfrac{s}{2}, \dfrac{1}{2}\right) & \text{if } s < 1 \\ \left(\dfrac{1}{2}, \dfrac{s}{2}\right) & \text{if } s \geq 1 \end{cases} \tag{A.8}$$

$$L_0 = L(h_0) = \frac{1}{2}\frac{s - t + 2h_0(t-1)}{s-1}.$$

Note that $h_0, L_0, x_4, y_1 > 0$. Recall finally that for $Q_{s,t}$ we assume that $s, t > 0, s + t > 1, s \neq 1$. Write the points of tangency of E_0 with $Q_{s,t}$ as $\zeta_1 = (0, y_1), \zeta_2 = (x_2, y_2)$, where $y_2 = 1 + \dfrac{t-1}{s}x_2, \zeta_3 = (x_3, y_3)$, where $y_3 = \dfrac{t}{s-1}(x_3 - 1)$, and $\zeta_4 = (x_4, 0)$. $\overleftrightarrow{\zeta_1\zeta_3}$ has equation $y = y_1 + \dfrac{y_3 - y_1}{x_3}x$ and $\overleftrightarrow{\zeta_2\zeta_4}$ has equation $y = -\dfrac{y_2}{x_4 - x_2}(x - x_4)$. By Lemma A.10, with $Q = Q_{s,t}$, $\overleftrightarrow{\zeta_1\zeta_3}$ and $\overleftrightarrow{\zeta_2\zeta_4}$ intersect at $\left(\dfrac{s}{s+t}, \dfrac{t}{s+t}\right)$, which implies that

$$y_1 + \frac{y_3 - y_1}{x_3}\frac{s}{s+t} = \frac{t}{s+t}, \tag{A.9}$$

$$-\frac{y_2}{x_4 - x_2}\left(\frac{s}{s+t} - x_4\right) = \frac{t}{s+t}.$$

Solving the second equation in (A.9) for x_2 using $y_2 = 1 + \dfrac{t-1}{s}x_2$ yields

$$x_2 = \frac{s^2(1 - x_4)}{(t-1)(s+t)x_4 + s}, \tag{A.10}$$

which then implies that

$$y_2 = t\frac{(t-1)x_4 + s}{(t-1)(s+t)x_4 + s}. \tag{A.11}$$

We also assume now that $s \neq t$ and that $t \neq 1$. Each of those cases can easily be handled separately. Now apply Lemma A.9 to the following pairs of tangency points.

- $A = \zeta_1$ and $B = \zeta_4$: Then $P = (0, 0)$ and the line, $L_{1,4}$, joining P to the center (h_0, L_0) has equation $y = \dfrac{L_0}{h_0}x$. Since $L_{1,4}$ intersects the chord $\overline{\zeta_1\zeta_4}$ at $P_{1,4} = \left(\dfrac{x_4 y_1 h_0}{x_4 L_0 + y_1 h_0}, \dfrac{x_4 y_1 L_0}{x_4 L_0 + y_1 h_0}\right)$, it follows that $P_{1,4} = \dfrac{1}{2}(x_4, y_1)$.

Equating components gives $\dfrac{y_1 h_0}{x_4 L_0 + y_1 h_0} = \dfrac{1}{2}$ and $\dfrac{x_4 L_0}{x_4 L_0 + y_1 h_0} = \dfrac{1}{2} \Rightarrow$

$$\frac{y_1 h_0}{x_4 L_0 + y_1 h_0} = \frac{x_4 L_0}{x_4 L_0 + y_1 h_0} \Rightarrow \frac{y_1}{x_4} = \frac{L_0}{h_0} \Rightarrow$$

$$y_1 = \frac{x_4\,(s - t + 2h_0 t - 2h_0)}{2\,(s - 1)\,h_0}. \tag{A.12}$$

- $A = \zeta_2$ and $B = \zeta_4$: Then $P = \left(-\dfrac{s}{t-1}, 0\right)$ and the line, $L_{2,4}$, joining P to the center (h_0, L_0) has equation $y = \dfrac{L_0}{h_0 + \dfrac{s}{t-1}}\left(x + \dfrac{s}{t-1}\right)$. Let

$$V_{2,4} = (t-1)(x_4 - x_2)L_0 + y_2(t-1)h_0 + y_2 s.$$

Note that $V_{2,4} = 0 \Rightarrow \dfrac{L_0}{h_0 + \dfrac{s}{t-1}} = \dfrac{y_2}{x_2 - x_4} \Rightarrow$ slope of $L_{2,4} =$ slope of the chord $\overline{\zeta_2 \zeta_4}$. But that would violate Lemma A.9 and thus $V_{2,4} \neq 0$. Since $L_{2,4}$ intersects the chord $\overline{\zeta_2 \zeta_4}$ at $P_{2,4} = \dfrac{((t-1)h_0 + s)x_4 y_2 - s(x_4 - x_2)L_0,\, y_2 L_0((t-1)x_4 + s)}{V_{2,4}}$, it follows that $P_{2,4} = \dfrac{1}{2}((x_2 + x_4), y_2)$. Equating components gives

$$((t-1)h_0 + s)x_4 y_2 - s(x_4 - x_2)L_0 = \frac{(x_2 + x_4)V_{2,4}}{2}$$

$$\text{and } ((t-1)x_4 + s)L_0 = \frac{V_{2,4}}{2},$$

which implies that $((t-1)h_0 + s)x_4 y_2 - s(x_4 - x_2)L_0 = (x_2 + x_4)L_0((t-1)x_4 + s)$. Substituting $L_0 = \dfrac{1}{2}\dfrac{s - t + 2h_0(t-1)}{s-1}$ and using (A.10) and (A.11) yields

$$h_0 = \frac{1}{2}\frac{(t-s)x_4 + s}{(t-1)x_4 + 1}. \tag{A.13}$$

Substituting for h_0 into (A.12) using (A.13) yields

$$y_1 = \frac{t x_4}{(t-s)x_4 + s}.$$

Solving the first equation in (A.9) for x_3 and using $y_3 = \dfrac{t}{s-1}(x_3 - 1)$ yields $x_3 = \dfrac{(t-1)x_4 + s}{(s + t - 2)x_4 + 1}$ and $y_3 = \dfrac{t(1 - x_4)}{(s + t - 2)x_4 + 1}$.

Substituting q for x_4 yields the tangency points $\zeta_1 - \zeta_4$. Using $L_0 = \dfrac{1}{2}\dfrac{s - t + 2h_0(t-1)}{s-1}$ yields

$$L_0 = \frac{1}{2}\frac{t}{(t-1)q + 1}. \tag{A.14}$$

Since $C_q = (h_0, L_0)$, that completes the proof of Proposition 1.1(ii). To prove Proposition 1.1(i), since E_0 has center $C_q = (h_0, L_0)$ by Theorem 1.1, one may write the equation of E_0 in the form

$$(x - h_0)^2 + B(x - h_0)(y - L_0) + C(y - L_0)^2 + F = 0. \qquad (A.15)$$

for some constants $B, C,$ and F. ∎

Differentiating (A.15) with respect to x yields $\dfrac{dy}{dx} = -\dfrac{2(x - h_0) + B(y - L_0)}{B(x - h_0) + 2C(y - L_0)}$.

We now let $r = \dfrac{qt}{(t - s)q + s}$. Since $\dfrac{dy}{dx}\big|_{\zeta_1} = \infty$ and $\dfrac{dy}{dx}\big|_{\zeta_4} = 0$, plugging in $\zeta_1 = (0, r)$ and $\zeta_4 = (q, 0)$ into $\dfrac{dy}{dx}$ yields

$$q - h_0 = \frac{BL_0}{2},$$

$$r - L_0 = \frac{Bh_0}{2C}.$$

Thus $B = \dfrac{2(q - h_0)}{L_0}$, and using (A.13) and (A.14) and simplifying gives

$$B = \frac{4q^2(t - 1)t + 2qt(s - t + 2) - 2st}{t^2}. \quad C = \frac{1}{2}\frac{Bh_0}{r - L_0},$$ which simplifies

to $\dfrac{((1 - q)s + qt)^2}{t^2}$. To obtain the constant term in (1.5), use $(q - h_0)^2$ $- BL_0(q - h_0) + CL_0^2 + F = 0 \Rightarrow F = BL_0(q - h_0) - (q - h_0)^2 - CL_0^2$, which simplifies to $\dfrac{q(q - 1)((t - 1)q + s)}{(t - 1)q + 1}$. Thus the constant term is $h_0^2 +$

$$Bh_0L_0 + CL_0^2 + F = \frac{q(qt - qs + s)}{(t - 1)q + 1} + \frac{q(q - 1)((t - 1)q + s)}{(t - 1)q + 1} = q^2.$$ Expand-

ing in (A.15) and then multiplying through by t^2 gives the coefficients of $xy, y^2, x,$ and y, respectively, in (1.5). Thus we have shown that if E_0 is an ellipse inscribed in $Q_{s,t}$, then the equation of E_0 is given by (1.5). Conversely, if E_0 is a conic with equation given by (1.5), then using basic calculus techniques it is easy to show that the lines $L_1 - L_4$ are each tangent lines to E_0, and that $\zeta_1 - \zeta_4$ are the points of tangency. Substituting for $A(q)$ thru $F(q)$ using (1.6) gives $\Delta = 16t^2(1 - q)q((t - 1)q + 1)((t - 1)q + s)$ and $\delta = 16t^4q^2(q - 1)^2((t - 1)q + s)^2$, each of which is positive by Lemma 1.2, and the fact that $q_0 \in J$ and $(s, t) \in G$. Since $A(q) > 0$ and $C(q) > 0$, by Lemma A.1 in the Appendix, E_0 is an ellipse. Using Lemma 1.2, it is not hard to show that $\zeta_j \in S_j, j = 1, 2, 3, 4$. Thus E_0 is an ellipse inscribed in $Q_{s,t}$, which completes the proof of Proposition 1.1(ii).

A.2.2   Proposition 1.2

Proof. Suppose that Q is **not** a parallelogram. Then by affine invariance we may assume that $Q = Q_{s,t}$, the quadrilateral given in Lemma 1.1. By

Proposition 1.1(i), E_1 and E_2 have centers $C_{q_1} = \dfrac{1}{2}\dfrac{1}{(t-1)q_1+1}((t-s)q_1+s,t)$

and $C_{q_2} = \dfrac{1}{2}\dfrac{1}{(t-1)q_2+1}((t-s)q_2+s,t)$, respectively. If $f(q) = \dfrac{t}{(t-1)q+1}$,

then $f'(q) = \dfrac{t(1-t)}{(qt-q+1)^2} > 0 \Rightarrow f$ is increasing. Thus $\dfrac{1}{2}\dfrac{t}{(t-1)q_1+1} \neq$

$\dfrac{1}{2}\dfrac{t}{(t-1)q_2+1} \Rightarrow C_{q_1} \neq C_{q_2}$. $\blacksquare$

A.2.3   Proposition 1.3

Proof. We could prove Proposition 1.3 using the same method used to prove Proposition 1.1. However, it is much easier to use the result of Proposition 1.1 along with a linear transformation from $Q_{s,t}$ to $Q_{S,T,v,w}$–note that S and T are used below instead of s and t since the corresponding values are not necessarily equal.

Let

$$\mathbb{F}(x,y) = (vx, wx+y). \tag{A.16}$$

By looking at the vertices, it follows easily that $\mathbb{F}(Q_{s,t}) = Q_{S,T,v,w}$, where $S = vs$ and $T = ws+t$. Note that $s = \dfrac{S}{v}$ and $t = T - \dfrac{wS}{v}$. First we want to show that if (1.10), (1.12), and (1.13) hold (using S and T), then $(s,t) \in G = \{(s,t) : s,t > 0, s+t > 1, s \neq 1\}$, which is necessary in order to apply Proposition 1.1. Let $f_1 = v(T-1) + (1-w)S$ and $f_2 = vT - wS$.

(1.10): $S > 0 \Rightarrow vs > 0 \Rightarrow s > 0$; we don't need to look at $t > w$

(1.12): $f_1 > 0 \Rightarrow v(T-1) + (1-w)S > 0 \Rightarrow \dfrac{S}{v} + T - \dfrac{wS}{v} > 1 \Rightarrow s+t > 1$

$f_2 > 0 \Rightarrow T - \dfrac{wS}{v} > 0 \Rightarrow t > 0$

(1.13): $S \neq v \Rightarrow \dfrac{S}{v} \neq 1 \Rightarrow s \neq 1$; we don't need to look at $f_1 \neq S$

Suppose now that E_0 is an ellipse inscribed in $Q_{S,T,v,w}$ and let $E_1 = \mathbb{F}^{-1}(E_0)$. Since E_1 is an ellipse inscribed in $Q_{s,t}$, the equation of E_1 is given by (1.5). Let $r = \dfrac{qt}{(t-s)q+s}$. Note that (1.5) is parameterized by q, where E_1 is tangent at $(q,0)$, whereas equation (1.15) is parameterized by r, where E_0 is tangent at $(0,r)$ (the reason for doing this is to simplify the proof of Theorem 1.4 somewhat). Since $\mathbb{F}^{-1}(x,y) = \left(\dfrac{x}{v}, -\dfrac{wx}{v} + y\right)$, the equation of E_0 is obtained from (1.5) by replacing x by $\dfrac{x}{v}$, replacing y by $-\dfrac{wx}{v} + y$, replacing s by $\dfrac{S}{v}$, replacing t by $T - \dfrac{wS}{v}$, and finally by replacing q by $\dfrac{rs}{r(s-t)+t}$. After clearing denominators and some more simplification, one obtains the equation in (1.15), with $s = S$ and $t = T$, which finishes the proof of (i). To prove (ii)–E_1 is tangent to $Q_{s,t}$ at the points $\zeta_1 - \zeta_4$ given in Proposition 1.1(ii). Applying $\mathbb{F}$ to $\zeta_1 - \zeta_4$ and simplifying yields

$$\mathbb{F}(\zeta_1) = \left(0, \frac{qt}{(t-s)q+s}\right) = (0,r),$$

$$\mathbb{F}(\zeta_2) = \frac{(v(1-q)s^2, w(1-q)s^2 + t(s+q(t-1)))}{(t-1)(s+t)q+s} = \frac{(Sv(1-r), f_1r + vT(1-r))}{f_1r + v(1-r)},$$

$$\mathbb{F}(\zeta_3) = \frac{(v(s+q(t-1)), w(s+q(t-1)) + (1-q)t)}{(s+t-2)q+1} =$$

$$\frac{((f_2 - (v-S)r)vS, wSf_1r + vTf_2(1-r))}{(S-v)(v+f_1)r + vf_2}, \text{ and } \mathbb{F}(\zeta_4) = (vq, wq) =$$

$$\frac{rS(v,w)}{Sr + f_2(1-r)}. \text{ Now applying } \mathbb{F} \text{ to the center in Proposition 1.1(i) gives}$$

$$\mathbb{F}\left(\frac{((t-s)q+s, t)}{2((t-1)q+1)}\right) = \frac{(v((t-s)q+s), w((t-s)q+s) + t)}{2((t-1)q+1)} =$$

$$\frac{(Sv, rS(1+w) - Tv(r-1))}{2((S-v)r+v)}. \text{ That yields Proposition 1.3(ii), with } s = S \text{ and}$$

$t = T$.

Conversely, suppose that E_0 is an ellipse with equation given by (1.15). One could again use Proposition 1.1, but it is easier and more direct to do the

following: The slope at any point $(x.y) \in E_0$ is given by $Z(x,y) = -\dfrac{\dfrac{\partial\psi(x,y)}{\partial x}}{\dfrac{\partial\psi(x,y)}{\partial y}}$;

First, it is easy to show that E_0 passes thru the four points $\zeta_1 - \zeta_4$ by showing that $\psi(\zeta_j) = 0, j = 1, ..., 4$; One can then show that the numerator of $Z(\zeta_1)$ is nonzero, while the denominator of $Z(\zeta_1)$ equals 0; Hence the tangent at ζ_1 is vertical, which matches the slope of L_1; In addition, $Z(\zeta_2) = \dfrac{t-1}{s} = $ slope of L_2, $Z(\zeta_3) = \dfrac{t-w}{s-v} = $ slope of L_3, and $Z(\zeta_4) = \dfrac{w}{v} = $ slope of L_4; Thus E_0 is tangent to Q at the points $\zeta_1 - \zeta_4$; It also follows easily by Lemma A.1 that E_0 is an ellipse. ∎

A.2.4 Lemma 2.1

Proof. The sides of Q, going clockwise, are given by $S_1 = \overline{A_1A_2}, S_2 = \overline{A_2A_3}, S_3 = \overline{A_3A_4}$, and $S_4 = \overline{A_4A_1}$. The lines containing the sides, going clockwise, have equations $x = -1, y = 1, x = 1$, and $y = -1$. Now suppose that E_0 is an ellipse inscribed in Q and suppose that E_0 is tangent to Q at the points $\zeta_1 = (-1, v) \in S_1, \zeta_2 = (u, 1) \in S_2, \zeta_3 = (1, t) \in S_3$, and $\zeta_4 = (s, -1) \in S_4$, where $-1 < s, t, u, v < 1$.

- Apply Lemma A.9 with $A = \zeta_1$ and $B = \zeta_2$: Then $P = (-1, 1)$ and the line, $L_{1,2}$, joining P to the center $(0, 0)$ has equation $y = -x$. The chord

$\overline{\zeta_1\zeta_2}$ has equation $y = 1 + \dfrac{1-v}{u+1}(x-u)$ and it follows that $L_{1,2}$ intersects $\overline{\zeta_1\zeta_2}$ at $P_{1,2} = \left(-\dfrac{1+uv}{u-v+2}, \dfrac{1+uv}{u-v+2}\right)$. Since $L_{1,2}$ bisects $\overline{\zeta_1\zeta_2}$, we also have $P_{1,2} = \dfrac{1}{2}(u-1, v+1)$. Hence $-\dfrac{1+uv}{u-v+2} = \dfrac{1}{2}(u-1)$ and $\dfrac{1+uv}{u-v+2} = \dfrac{1}{2}(v+1)$. Either equation implies that $v = -u$.

- Apply Lemma A.9 with $A = \zeta_2$ and $B = \zeta_3$: Then $P = (1,1)$ and the line, $L_{2,3}$, joining P to the center $(0,0)$ has equation $y = x$. The chord $\overline{\zeta_2\zeta_3}$ has equation $y = t + \dfrac{t-1}{1-u}(x-1)$ and it follows that $L_{2,3}$ intersects $\overline{\zeta_2\zeta_3}$ at $P_{2,3} = \left(\dfrac{tu-1}{u+t-2}, \dfrac{tu-1}{u+t-2}\right)$. Since $L_{2,3}$ bisects $\overline{\zeta_2\zeta_3}$, it follows that $P_{2,3} = \dfrac{1}{2}(u+1, t+1)$. Hence $\dfrac{tu-1}{u+t-2} = \dfrac{1}{2}(u+1)$ and $\dfrac{tu-1}{u+t-2} = \dfrac{1}{2}(t+1)$. Either equation implies that $t = u$.

- Finally, apply Lemma A.9 with $A = \zeta_1$ and $B = \zeta_4$: Then $P = (-1,-1)$ and the line, $L_{1,4}$, joining P to the center $(0,0)$ has equation $y = x$. The chord $\overline{\zeta_1\zeta_4}$ has equation $y = -1 - \dfrac{v+1}{s+1}(x-s)$ and it follows that $L_{1,4}$ intersects $\overline{\zeta_1\zeta_4}$ at $P_{1,4} = \left(\dfrac{sv-1}{s+v+2}, \dfrac{sv-1}{s+v+2}\right)$. Since $L_{1,4}$ bisects $\overline{\zeta_1\zeta_4}$, it follows that $P_{1,4} = \dfrac{1}{2}(s-1, v-1)$. Hence $\dfrac{sv-1}{s+v+2} = \dfrac{1}{2}(s-1)$ and $\dfrac{sv-1}{s+v+2} = \dfrac{1}{2}(v-1)$. Either equation implies that $s = v$. That proves Lemma 2.1(ii).

To prove (i): Since the center of E_0 equals the intersection point, $(0,0)$, of the diagonals of Q, we may assume that the equation of E_0 has the form $Ax^2 + Bxy + Cy^2 + F = 0$ with $A, C > 0$. Plugging in ζ_2 yields $Au^2 + Bu + C + F = 0$ and plugging in ζ_3 yields $A + Bu + Cu^2 + F = 0$. Subtracting gives $Au^2 - A + C - Cu^2 = 0 \Rightarrow (A-C)u^2 - (A-C) = 0$. Since $u \neq \pm 1$, this implies that $C = A$. Differentiating with respect to x yields $\dfrac{dy}{dx} = -\dfrac{2Ax + By}{Bx + 2Cy}$. Plugging in ζ_2 and matching slopes yields $2A(u) + B(1) = 0 \Rightarrow B = -2Au$. That gives $F = -(Au^2 + Bu + C) = Au^2 - A$. The equation then becomes $Ax^2 - 2Auxy + Ay^2 + Au^2 - A = 0$. Dividing thru by A yields (2.1). Conversely, if E_0 is a conic section with equation given by (2.1), then using basic calculus techniques it is easy to show that E_0 is inscribed in Q. It also follows easily by Lemma A.1 that E_0 is an ellipse. ∎

A.2.5 Lemma 2.4

Proof. Denote the vertices of Q, going clockwise and starting with the lower left corner vertex, by $A_j = (x_j, y_j), j = 1 - 4$. Note that $\overleftrightarrow{A_1 A_2} \parallel \overleftrightarrow{A_3 A_4}$ and $\overleftrightarrow{A_2 A_3} \parallel \overleftrightarrow{A_1 A_4}$. Using a translation, we may assume that Q has center $= (0,0)$, which implies that $x_1, x_2, y_1, y_4 < 0 < x_3, x_4, y_2, y_3$. Using a rotation, we may also assume that Q has **horizontal** parallel sides, which implies that $y_3 = y_2$ and $y_4 = y_1$. Finally, using a reflection thru the x-axis, we may assume that $x_4 < x_3$. The diagonal lines of Q are $\overleftrightarrow{A_1 A_3}$ and $\overleftrightarrow{A_2 A_4}$, which have equations
$$y = y_1 + \frac{y_2 - y_1}{x_3 - x_1}(x - x_1) \text{ and } y = y_2 + \frac{y_1 - y_2}{x_4 - x_2}(x - x_2).$$ $\overleftrightarrow{A_1 A_3}$ and $\overleftrightarrow{A_2 A_4}$
intersect at $IP = \left(\dfrac{x_1 x_2 - x_3 x_4}{x_1 + x_2 - x_3 - x_4}, \dfrac{(x_2 - x_3)y_1 + (x_1 - x_4)y_2}{x_1 + x_2 - x_3 - x_4} \right)$. Since the center of Q is $(0,0)$, the diagonal lines must intersect at $(0,0)$, which implies that $x_1 x_2 - x_3 x_4 = 0$ and $(x_2 - x_3)y_1 + (x_1 - x_4)y_2 = 0$.

We consider two cases.

Case 1: Q is not a rectangle. Then $x_1 \neq x_2$ and $x_3 \neq x_4$. $\overleftrightarrow{A_1 A_2} \parallel \overleftrightarrow{A_3 A_4} \Rightarrow$
$\dfrac{y_2 - y_1}{x_2 - x_1} = \dfrac{y_4 - y_3}{x_4 - x_3} \Rightarrow \dfrac{y_2 - y_1}{x_2 - x_1} = \dfrac{y_1 - y_2}{x_4 - x_3} \Rightarrow x_4 - x_3 = x_1 - x_2 \Rightarrow$
$x_1 - x_4 = x_2 - x_3$. Thus $(x_2 - x_3)y_1 + (x_1 - x_4)y_2 = (x_2 - x_3)(y_1 + y_2) = 0 \Rightarrow y_1 = -y_2$. Then $y_3 = -y_4$ as well. $x_4 = \dfrac{x_1 x_2}{x_3}$ and $x_4 = x_1 - x_2 + x_3$
imply that $\dfrac{x_1 x_2}{x_3} = x_1 - x_2 + x_3 \Rightarrow x_3^2 + (x_1 - x_2)x_3 - x_1 x_2 = 0 \Rightarrow$
$(x_1 + x_3)(x_3 - x_2) = 0 \Rightarrow x_1 = -x_3$. Thus $x_2 = -x_4$ as well.

Case 2: Q is a rectangle. Then $x_1 = x_2$ and $x_3 = x_4$ and thus $x_1 x_2 - x_3 x_4 = 0$ implies that $x_1^2 = x_3^2 \Rightarrow x_1 = -x_3$ since $x_1 \neq x_3$. Then $x_2 = -x_4$ as well.

$(x_2 - x_3)y_1 + (x_1 - x_4)y_2 = 0$ then becomes, after some simplification, $y_1 + y_2 = 0$ and thus $y_1 = -y_2$. Then $y_3 = -y_4$ as well.

In either case, let $l = \dfrac{1}{2}(x_3 + x_4), d = \dfrac{1}{2}(x_3 - x_4)$, and $k = y_3$, which implies that $y_4 = -k$ and thus $y_2 = k$ and $y_1 = -k$. Also $l, k > 0, 0 \leq d < l, x_3 = l + d$, and $x_4 = l - d$, which implies that $x_1 = -l - d$ and $x_2 = -l + d$. ∎

References

[1] John Clifford and Michael Lachance, Quartic Coincidences and the Singular Value Decomposition, *Mathematics Magazine*, December, 2013, 340–349.

[2] Hayes, Copeland, Gfrerrer, Zsombor-Murray, Largest Area Ellipse Inscribing an Arbitrary Convex Quadrangle, In: Uhl, T. (eds) Advances in Mechanism and Machine Science. IFToMM WC 2019. *Mechanisms and Machine Science*, vol 73. Springer, Cham. https://doi.org/10.1007/978-3-030-20131-9_24

[3] Zachary A. Copeland, A Completely Generalized and Algorithmic Approach to Identifying the Maximum Area Inscribing Ellipse of an Asymmetric Convex Quadrangle, Masters Thesis, Ottawa-Carleton Institute for Mechanical and Aerospace Engineering, 2017.

[4] Jason P. Byrne, Shane A. Maloney, R. T. James McAteer, Jose M. Refojo, and Peter T. Gallagher, "Propagation of an Earth-directed coronal mass ejection in three dimensions", *Nature Communications* 2010, Vol. 1, Article Number 74.

[5] Alexander Karmazin and Karlheinz Spindler, "Identification of the Contents of a Bottle From Two X-Ray Views", *Advances in Mathematical and Computational Methods*, p. 313–317.

[6] Ping Yao, Jun Yu, Wenjie Wu and Lu Chen, "3D Surface Reconstruction Based on Binocular Vision", *Proceedings of 2014 IEEE International Conference on Mechatronics and Automation*, August 3-6, Tianjin, China.

[7] K. Yadav, S. M. Ishtiaque1, S. D. Joshi, and J. K. Chatterjee "Diametric unevenness and fault classification of yarn using newly developed diametric fault system", *Fibers and Polymers* 2017, Vol.18, No. 10, 2018–2033.

[8] Nidish Narayanaa Balaji, Multi-Scale Modeling in Bolted Interfaces, Master's Thesis, Department of Mechanical Engineering, William Marsh Rice University, 2019, pp. 1–101.

[9] Nidish Narayanaa Balajia, Wei Chenb, Matthew R.W. Brake, Traction-based multi-scale nonlinear dynamic modeling of bolted joints: Formulation, application, and trends in micro-scale interface evolution, *Mechanical Systems and Signal Processing* 139 (2020), pp. 1–32.

[10] E.J. Garboczia, K.A. Ridingb, Mohammadreza Mirzahosseinic, "Particle shape effects on particle size measurement for crushed waste glass", *Advanced Powder Technology* 28(2), December, 2016.

[11] Michael Demeyere and Christian Eugene, Measurement of Cylindrical Objects by Laser Telemetry in an Ambulatory Context, *IEEE Instrumentation and Measurement*, Technology Conference, Anchorage, AK, USA, 21-23 May 2002, pp. 1–6.

[12] Michael Demeyere and Christian Eugene, "Measurement of Cylindrical Objects by Laser Telemetry in a Robotic Environment", *European Journal of Mechanical and Environmental Engineering*, Vol. 47, No. 4, p. 209–213.

[13] Alan Horwitz, "Ellipses of maximal area and of minimal eccentricity inscribed in a convex quadrilateral", *Australian Journal of Mathematical Analysis and Applications*, 2(2005), 1–12.

[14] Alan Horwitz, Ellipses Inscribed in Parallelograms, *Australian Journal of Mathematical Analysis and Applications*, Volume 9, Issue 1(2012), 1–12.

[15] Alan Horwitz, Dynamics of ellipses inscribed in quadrilaterals, *IOSR Journal of Mathematics*, Volume 15, Issue 5, Series-4 (Sep-Oct 2019).

[16] Alan Horwitz, Ellipses of minimal area and of minimal eccentricity circumscribed about a convex quadrilateral, *Australian Journal of Mathematical Analysis and Applications*, Volume 7, Issue 1, Article 8(2010).

[17] Alan Horwitz, An area inequality for ellipses inscribed in quadrilaterals, *Journal of Mathematical Inequalities*, 4(2010), Issue 3, 431–443.

[18] Alan Horwitz, Midpoint Diagonal Quadrilaterals, *International Journal of Geometry*, Vol. 11 (2022), No. 3, July, 102–123.

[19] Alan Horwitz, When is an ellipse inscribed in a quadrilateral tangent at the midpoint of two or more sides?", *IOSR Journal of Mathematics*, Volume 16, Issue 1 Ser. I (Jan – Feb 2020), 61–67.

[20] G. D. Chakerian, A Distorted View of Geometry, MAA, Mathematical Plums, Washington, DC, 1979, 130–150.

[21] Heinrich Dörrie, "Newton's Ellipse Problem", *100 Great problems of Mathematics, Their History and Solution*, Dover Publications, Inc. New York, 1965, 217–219.

[22] M. John D. Hayes, Maximum Area Ellipses Inscribing Specific Quadrilaterals, *Proceedings of The Canadian Society for Mechanical Engineering International Congress 2016*, 2016 CCToMM M3 Symposium, June 26–29, 2016, Kelowna, British Columbia, Canada.

[23] http://en.wikipedia.org/wiki/Tangential_quadrilateral

[24] https://en.m.wikipedia.org/wiki/Conic_section

[25] Morris Marden, The Location of the Zeros of the Derivative of a Polynomial, *American Mathematical Monthly*, Vol. 42, No. 5 (May, 1935), pp. 277–286.

[26] Thirty-Third William Lowell Putnam Mathematical Competition, 1972, Problem A4: Show that a circle inscribed in a square has a larger perimeter than any other ellipse inscribed in the square, *American Mathematical Monthly*, Vol. 80, No. 10 (Dec., 1973).

[27] D. Minda and S. Phelps, Triangles, Ellipses, and Cubic Polynomials, *American Mathematical Monthly* 115(2008), 679–689.

[28] Constantin P. Niculescu, A New Look at Newton's Inequalities, *JIPAM*, Volume 1, Issue 2, Article 17, 2000.

[29] Martin Josefsson, Calculations Concerning the Tangent Lengths and Tangency Chords of a Tangential Quadrilateral, *Forum Geometricorum*, Volume 10 (2010) 119–130.

[30] https://en.wikipedia.org/wiki/Tangential_quadrilateral

[31] Martin Josefsson, Great compilation of characterizations of squares, *International Journal of Geometry*, Vol. 12 (2023), No. 3, 13–37.

[32] Martin Josefsson, Properties of bisect-diagonal quadrilaterals, *The Mathematical Gazette*, Vol. 101(2017), No. 551, 214–226.

[33] Michael de Villiers, Some more properties of the bisect-diagonal quadrilateral, *The Mathematical Gazette*, Vol. 105(2021), No. 564, 474–480.

[34] Heinrich Dörrie: "Steiner's Ellipse Problem", *100 Great Problems of Elementary Mathematics*, Dover, New York, 1965, 378–381.

[35] Alan Horwitz, Dynamics of ellipses inscribed in triangles, *Journal of Science, Technology and Environment*, Volume 5, Issue 1 (2016), 1–21.

[36] Heinrich Dörrie, "The most nearly circular ellipse circumscribing a quadrilateral", *100 Great Problems of Mathematics, Their History and Solution*, Dover Publications, Inc. New York, 1965, 231–236.

[37] Jia Hui Li, Zhuo Qun Wang, Yi Xi Shen, and Zhong Yuan Dai, Does any convex quadrilateral have circumscribed ellipses?, *Opten Math.* 2017. 15:1463–1476.

[38] B. V. Rublev and Yu. I. Petunin, "Minimum–Area Ellipse Containing a Finite Set of Points. 1, *Ukrainian Mathematical Journal*, Vol. 50, No. 7, 1998, 1115–1124.

[39] B. V. Rublev and Yu. I. Petunin, "Minimum–Area Ellipse Containing a Finite Set of Points. 2, *Ukrainian Mathematical Journal*, Vol. 50, No. 8, 1998, 1253–1261.

[40] Alsina, Claudi; Nelsen, Roger (2009), "4.3 Cyclic, tangential, and bicentric quadrilaterals", *When Less is More: Visualizing Basic Inequalities*, Mathematical Association of America, p. 64, Dolciani Mathematical Expositions 36, Underwood Dudley, Editor, Washington, DC, ISBN 978-0-88385-342-9

[41] https://mathworld.wolfram.com/Ellipse.html

[42] Mohamed Ali Said, "Calibration of an Ellipse's Algebraic Equation and Direct Determination of its Parameters", *Acta Mathematica Academiae Paedagogiace Nyíregyháziensis,* Vol.19, No. 2 (2003), 221–225.

[43] David A. Brannan, Matthew F. Esplen, and Jeremy J. Gray, *Geometry*, Cambridge University Press, Cambridge, UK, 1999.

[44] C. Bond, "A New Algorithm for Scan Conversion of a General Ellipse", preprint, http://www.crbond.com/papers/ell_alg.pdf

Index

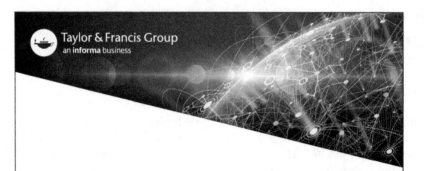

Printed in the United States
by Baker & Taylor Publisher Services